COMPUTER SCIENCE, TECHNOLOGY AND APPLICATIONS

COMPUTATIONAL MECHANICS (CM)

APPLICATIONS AND DEVELOPMENTS

Computer Science, Technology and Applications

Additional books in this series can be found on Nova's website under the Series tab.

Additional e-books in this series can be found on Nova's website under the eBooks tab.

COMPUTER SCIENCE, TECHNOLOGY AND APPLICATIONS

COMPUTATIONAL MECHANICS (CM)

APPLICATIONS AND DEVELOPMENTS

JACOB YUEN
EDITOR

nova
science publishers
New York

Library of Congress Cataloging-in-Publication Data

ISBN: 978-1-53613-672-2

Published by Nova Science Publishers, Inc. † New York

CONTENTS

PREFACE

In this collection, a model based on the modified strain gradient theory for the free vibration analysis of microbeams considering the effect of rotatory inertia and of shear deformation is presented. An analysis of the effects of boundary supports of the microbeams on the natural non-dimensional frequencies is performed. The results are obtained using the Ritz method with sets of admissible functions that can best model the microbeam boundary conditions. The following work employs a physically interpretable notation called strain gradient notation to study the sources and effects of locking. Such notation allows for the physical meanings of the polynomials coefficients to be identified early in the formulation procedure. Strain gradient notation guarantees that the sources of locking can be eliminated a-priori, rendering efficient finite element models. Another study is included in which the evolutionary process accounts for the threshold values for the Von Mises stress. Thus, the regions with lower Von Mises stresses intensity have removal condition whereas the higher Von Mises stresses regions have addition condition. This study aims to evaluate the reliability of structures optimized topologically as well as to determine the robustness of the solutions provided by the topology optimization schemes. In the closing work, the microwave heating of a liquid product inside a tank (reactor) with and without mechanical agitation was analyzed. A mathematical model was developed that includes the resolution of the microscopic energy balance in which a source term is used to consider the interaction between food and microwaves. From numerical simulations it was possible to evaluate the effect of the agitation rate, the position and number of the microwave-generators inside the cavity in the temperature profiles of the liquid products.

Chapter 1 - Beams of materials whose properties vary functionally along the axis (AFG) and the cross section varies gradually (tapered) are studied in the present investigation. The flexural behavior of the beam is described by the Timoshenko theory which takes into account the rotary inertia and shear deformation effects. General boundary conditions are considered, by means of translational and rotatory springs at both ends which allow us to represent all possible combinations of classical boundary conditions. The problems are solved by applying approximate methods: the well-known Rayleigh-Ritz method, the Generalized Differential Quadrature method and an approach by the Finite Element method are used. The objective of this study is twofold: from the point of view of structural mechanics to solve a difficult elastodynamic problem that has innumerable technological applications and from the point of view of computational mechanics the demonstration of the aptitude of classical numerical procedures to attack the problem as well as the possibility of comparing its accuracy. A variety of numerical examples are evaluated by combining boundary conditions, variations in cross section and material composition. Dynamic stiffening effect is achieved and discussed in some cases. The results are in good agreement with particular situations of the model, available in the scientific literature.

Chapter 2 - The paper presents a model based on the modified strain gradient theory for the free vibration analysis of isotropic microbeams. The analysis includes the effects of rotatory inertia and shear deformation on natural frequencies. It takes into account the dependence of the mechanical properties of the material on the size of the microbeams. The model is sensitive to the Poisson's ratio and its effect can be observed on the natural frequency coefficients. The numerical results developed in this paper include Timoshenko and Bernoulli – Euler microbeams considering different boundary conditions: simply supported–simply supported, clamped–free, clamped–simply supported, clamped–clamped, simply supported– free, free–free. The effects that boundary supports of the microbeams have on the natural non-dimensional frequencies are also studied. Numerical solutions are obtained using the Ritz method with sets of admissible functions that best model the boundary conditions of the microbeam. Additionally, results obtained with the classical theory and with the modified couple stress theory are also presented.

Chapter 3 - The displacement-based finite element method commonly adopts as approximating functions polynomial expansions whose coefficients are unknown. Such coefficients are carried along in the formulation procedure and their physical meanings, although usually unknown to the developer, permeate the entire formulation of the finite element. The use of incomplete and/or inconsistent polynomials will cause spurious terms to be part of the physical contents of the strain polynomials coefficients. Such spurious terms will be the sources of modeling errors which are known as membrane and shear locking. The deleterious effects of locking have been well described in the literature and their avoidance is paramount for efficient finite element solutions. In general, locking represents quantitative solution errors, meaning that convergence requires a great level of mesh refinement. Further, qualitative errors may arise, and refinement might not be able to overcome them. This work employs a physically interpretable notation called strain gradient notation to study the sources and effects of locking in plane elasticity elements. Such notation allows for the physical meanings of the polynomials coefficients to be identified early in the formulation procedure. This characteristic provides the developer with a clear understanding of the capabilities of the model, and allows for the sources of locking to be easily and precisely identified and removed. Such procedure is carried out without introducing spurious zero-energy modes. Conventional finite element formulations do not possess such capability. For instance, isoparametric finite elements require the use of reduced-order integration procedures, which are not very effective in certain cases as spurious zero-energy modes are introduced. Strain gradient notation guarantees that the sources of locking can be eliminated a-priori, rendering efficient finite element models.

Chapter 4 - The structural design couples attributes that are, normally, opposites: economy and safety. This type of coupling is handled properly by optimization techniques. Particularly, topology optimization approaches have supported the proposition of structural layouts accounting for the minimum amount of material. Among the topology optimization schemes presented in the literature, it is worth citing the Evolutionary Structural Optimization (ESO). In this approach, the material removal/addition processes are governed by a desired criterion. In the present study, the evolutionary process accounts for threshold values of the Von Mises stress. Therefore, Von Mises stress values smaller than the lower threshold lead to the removal condition. On the other hand, addition condition is observed for Von Mises stress values higher than the upper threshold. Moreover, the ESO provides the structural layout for the target volume. Because of the geometry complexity observed during the topology evolution procedure, numerical methods are required for assessing the stress state. In the present study, the Finite Element Method (FEM) is coupled to the ESO for topology optimization analyses. However, this coupling does not account for the inherent randomness that exist into all-mechanical problem. In this regard, this study proposes a numerical scheme for assessing the reliability of structures optimized topologically. The Monte Carlo simulation technique is utilized for determining the probability of structural failure. In addition to the safety aspects, this study clarifies the robustness of the solutions provided by the topology optimization schemes. Two applications are utilized to demonstrate the efficiency of the proposed model.

Chapter 5 - One of the most widely used technologies for food heating today is the use of microwaves. This procedure is applied daily at the domestic level but shows difficulties in industrial designs; one of them is the non-uniformity in the distribution of temperatures. This phenomenon is caused by an inhomogeneous distribution of the electromagnetic fields within the control volume, depending on the dielectric properties of the material and the geometry of the system. In this work, the microwave heating of a liquid product inside a tank (reactor) with and without mechanical agitation was analyzed. To this purpose, a mathematical model was developed that includes the resolution of the microscopic energy balance in which a source term is used to consider the interaction between food and microwaves. The electromagnetic distribution was obtained through the resolution of the Maxwell equations for the product and the oven-cavity. In the case of the stirred tank, the microscopic momentum balance was coupled to the non-stationary energy balance to solve the distribution of velocities within the tank. The Maxwell equations and the energy and momentum balances with the initial and boundary conditions were solved using the finite element method through the commercial software COMSOL Multiphysics. From numerical simulations it was possible to evaluate the effect of the agitation rate, the position and number of the microwave-generators inside the cavity in the temperature profiles of the liquid products.

In: Computational Mechanics (CM)
Editor: Jacob Yuen
ISBN: 978-1-53613-672-2

Chapter 1

FREE VIBRATIONS OF TAPERED AFG TIMOSHENKO BEAMS

Gonzalo J. Gilardi, Carlos A. Rossit*[*] and Diana V. Bambill
Departamento de Ingeniería,
Universidad Nacional del Sur (UNS), Bahía Blanca, Argentina
Consejo Nacional de Investigacioncs
Científicas y Técnicas, CONICET

ABSTRACT

Beams of materials whose properties vary functionally along the axis (AFG) and the cross section varies gradually (tapered) are studied in the present investigation. The flexural behavior of the beam is described by the Timoshenko theory which takes into account the rotary inertia and shear deformation effects. General boundary conditions are considered, by means of translational and rotatory springs at both ends which allow us to represent all possible combinations of classical boundary conditions.

* Corresponding Author Email: carossit@criba.edu.ar.

The problems are solved by applying approximate methods: the well-known Rayleigh-Ritz method, the Generalized Differential Quadrature method and an approach by the Finite Element method are used.

The objective of this study is twofold: from the point of view of structural mechanics to solve a difficult elastodynamic problem that has innumerable technological applications and from the point of view of computational mechanics the demonstration of the aptitude of classical numerical procedures to attack the problem as well as the possibility of comparing its accuracy.

A variety of numerical examples are evaluated by combining boundary conditions, variations in cross section and material composition. Dynamic stiffening effect is achieved and discussed in some cases. The results are in good agreement with particular situations of the model, available in the scientific literature.

Keywords: beam vibrations, AFG beam, Timoshenko beam, tapered beam

INTRODUCTION

As Mahamood et al. (2012) have established, there are numerous technological situations in which pure metals have no useful application. For example, if the material to be employed is required to have hardness and ductility at the same time, such a material is not known in nature. To solve the problem, two metals or one metal have been combined with a non-metallic material. An example of remarkable historical importance was the alloy between copper and tin to generate bronze. So much so that it has led to the so-called "Bronze Age" in the fourth millennium BC.

Since then, man has come a long way in order to improve the properties of the materials he has used as a support for his development. At present, composite materials, formed by two or more materials combined in the solid state, with physical and chemical properties different from the individual parental materials, are highly applicable. The most used are the reinforcements made of fibers of very resistant materials, oriented according to design requirements, in a ceramic matrix. Despite the enormous comparative advantages they have over conventional materials, the compounds can fail in extreme stress conditions through the process of "delamination," that is, the separation of the fibers from the matrix, which prevents the transmission of stresses, Wang (1983).

This problem was solved by Japanese researchers in the mid-eighties, when resorting to a new type of material, Niino et al. (1987), functionally gradual, called Functionally Graded Materials (FGM). It constitutes a class of advanced materials in which their properties vary gradually with some of their dimensions. In the FGM materials, the abrupt interfaces of the compounds are eliminated, replacing that interface with a graduated interface that produces a smooth transition between one material and the next. One of its distinctive features is the ability to adapt for a specific application.

These materials can be manufactured by varying the percentage content of two or more materials so that the new materials have the property with the desired gradation in established spatial directions. In this way, thermal stresses, residual stresses and stress concentration factors found in laminated compounds are reduced.

Under the favorable characteristics mentioned above, there have been many studies on the static and dynamic behavior of resistant structures of FGM. Initially, investigative work established great progress in the field of elasticity theory and the study of plates and shells built with FGM. Paradoxically progress in its application to beams (Functionally Graded Beams -FGB-) was much later.

Beams are probably the most widespread component of resistant structures. Consequently, enormous research effort has been developed about them. When they are in vibratory environments, it is important to know their dynamic parameters. In that case, the inertia effect magnifies the influence of the constituent materials. In recent years, it has become increasingly important to use advanced materials whose properties vary gradually in some of its dimensions (Functionally Graded Material -FGM). However, most of early work on FGB has considered the gradation of material properties in the thickness direction.

Sankar (2001) established a model of FG Euler-Bernoulli beam to treat a static problem in a simply supported beam. The corresponding case of free transverse vibrations was studied by Aydogdu and Taskin (2007). Additionally, Chabraborty et al. (2003) developed a finite element to study the thermoelastic behavior of FGB. In the works of Lu and Chen (2005) and Zhong and Yu (2007) complex analytical procedures are carried out to solve accurately FGB within the field of the Theory of Elasticity. All these studies were conducted based on the Euler-Bernoulli description of the beam deformation.

Far fewer researchers have considered the variation of material properties in the axial direction (Axially Functionally Graded - AFG). Probably because the problem becomes more complicated as variable coefficients appear in the governing differential equations, even for Bernoulli-Euler simplest theory.

Consequently, because of the mathematical difficulties in treatment, few analytical solutions have been obtained and for arbitrary specific gradients: For example Elishakoff and his colleagues -Elishakoff (2000), Elishakoff and Candan (2001), Caliò and Elishakoff (2004) and (2005), Elishakoff (2005), Elishakoff and Guede (2005) and Wu, Wang and Elishakoff (2005)-, who by means of the semi-inverse method, solved various particular cases of vibrating AFG beams.

Huang and Li (2010) solved the problem by transforming the governing equation with variable coefficients in a Fredholm integral equation. Alshorbgy et al. (2011) investigated the dynamic characteristics of non-uniform beams with axially or transversely in height gradation of the material by means of the finite element method. Shahba and Rajasekaran (2012) studied longitudinal and transverse free vibration and buckling of AFG Euler-Bernoulli beams using the differential transform element method (DTEM) and differential quadrature element method of lowest-order (DQEL). Hein and Feklistova (2011) investigated the vibration of non-uniform and AFG beams with various boundary conditions and varying cross-sections using Haar wavelets. Şimşek and co-workers (2011, 2012) studied the dynamic behavior of AFG beams under the action of a moving load and Chegenizadeh et al. (2014) studied statically and dynamically AFG beams embedded in an elastic medium.

AFG Bernoulli-Euler beams of varying cross section and different boundary conditions are studied also using various approaches: Ghazaryan et al. (2017) used the differential transform method, Chen et al. (2017) introduced a numerical method to transform the differential equation into a set of linear algebraic equations with the displacement function being expanded using Taylor series or Chebyshev polynomials and Xie et al. (2017) by means of a spectral collocation approach based on integrated polynomials.

Mentions must be made of the studies of AFG Bernoulli-Euler vibrating beams of varying cross section carrying attached masses, Rossit et al. (2017) using the Rayleigh-Ritz method and subsequently Nikolić (2017) by means of the rigid element method.

As it is known the classical Bernoulli-Euler beam theory adequately predicts the frequencies of vibration of lower modes of slender beams. Timoshenko theory, posed in 1921 (Timoshenko 1921, 1922), involves the effects of rotational inertia, shear deformation and their combined effects and provides more accurate results in the general study of beams and is mandatory in the case of high frequencies or non-slender beams. Remarks of historical character on the subject can be found in the paper by Elishakoff et al. (2015).

Among papers about AFG Timoshenko beams, mention must be made of the paper of Shahba et al. (2011) who studied free vibration and stability of AFG Timoshenko beams through a finite element approach. Huang et al. (2013) presented a new approach by introducing an auxiliary function, they changed the coupled governing equations with variable coefficients for the deflection and rotation to a single governing equation. He et al. (2013) improved the traditional beam element to consider the variable axial parameters which were formulated in terms of a power series. Tang et al. (2014) obtained closed form solutions for uniform AFG Timoshenko beams whose bending stiffness and distributed mass density are assumed to obey a unified exponential law. Rajasekaran and Norouzzadeh Tochaei (2014) investigated the free vibration analysis of AFG Timoshenko beams carried out through the differential transformation element method and the quadrature element method. They introduce an element-based differential method that significantly improves the accuracy of results. They also introduced a lower order differential quadrature element based on differential quadrature element method. They showed the accuracy of both method with several numerical examples.

Also, Sarkar and Ganguli (2014) founded closed form solutions for certain polynomial variations of the material mass density, elastic modulus and shear modulus, along the length of the beam. Gan et al. (2015) presented a finite element procedure for dynamic analysis of non-uniform AFG Timoshenko beams under multiple moving point loads. Bambill et al. (2015) studied AFG Timoshenko beams using the Generalized Differential Quadrature method. Huang et al. (2016) studied the buckling behaviors of axially functionally graded and non-uniform Timoshenko beams by means of the auxiliary function and power series. More recently, Zhao et al. (2017) introduced a new approach based on Chebyshev polynomials theory; Tudjono et al. (2017) derived exact shape functions for both non-uniform (non-prismatic section) and inhomogeneous (functionally graded material) Timoshenko beam element formulation explicitly and Gan et al. (2017) investigated the post-buckling analysis of planar beams and frames made of axially functionally graded material by using the finite element method. Recently, the effect of concentrated masses on Timoshenko AFG tapered beams is studied (Rossit et al., 2018).

In the present work the free vibrations of the AFG Timoshenko beams are analyzed. General boundary conditions are modeled by using translational and rotational springs at both ends, which allow us to represent all the possible combinations of classical boundary conditions, as well as elastic restraints.

As it was stated, except for a few specific situations, these problems are solved by applying approximate methods. In this development, the well-known Rayleigh-Ritz method, the Generalized Differential Quadrature method and an approach by the Finite Element method are used.

All the procedures show aptitude to tackle the problem through its satisfactory comparison with particular cases available in the literature.

Gradation of Material Properties

The inhomogeneous material, with gradient compositional variation of the constituents, varies in the longitudinal direction of the beam. Properties of AFG materials, like mass density ρ, Young's modulus E, shear modulus G, continuously vary in the axial direction.

In this case the following variations have been considered for a typical material property $P(\bar{x})$ such as ρ, E or G.

Symmetric Parabolic Variation of the Second Degree

$$P(\bar{x}) = P_a + 4(P_b - P_a)\left(1 - \frac{\bar{x}}{L}\right)\frac{\bar{x}}{L} \tag{1}$$

where P_a and P_b are properties of material "*a*" and material "*b*," respectively. They are the constituents of the inhomogeneous material of the beam. Note that for $\bar{x} = 0$ the entire section is of material "*a*," and for $\bar{x} = L/2$ the whole section is of material "*b*."

Linear or Exponential Non-Symmetric Variation

$$P(\bar{x}) = P_a + (P_b - P_a)\left(\frac{\bar{x}}{L}\right)^n \tag{2}$$

Note that for $\bar{x} = 0$ the entire section is of material "*a*," and for $\bar{x} = L$ the whole section is of material "*b*." The percentage content of material "*a*" along the beam increases as n increases. When $n = 1$ the composition changes linearly through the length L, while $n = 1/2$ or $n = 2$ corresponds to a quadratic distribution, and so on. In general, any value n outside the range (1/3, 3) is not desired (Nakamura et al., 2000) because such a functionally graded material would contain too much of one of the constituents. (When $n = 1/3$ or 3, one constituent has the 75% of the total AFG material).

DIFFERENTIAL SYSTEM

The differential equations that represent the flexural behavior of a AFG Timoshenko beam (Figure 1), were derived applying the Hamilton's principle and following the procedure of Banerjee (2001),

$$\left[\kappa AG\left(v'-\psi\right)\right]' - \rho A\ddot{v} = 0 \quad (3)$$

$$\left(EI\psi'\right)' + \kappa AG\left(v'-\theta\right) - \rho I\ddot{\psi} = 0 \quad (4)$$

where $v = v\left(\overline{x},t\right)$ is the flexural displacement of the beam neutral axis in the $\overline{y}$ direction and $\psi = \psi\left(\overline{x},t\right)$ is the cross-section rotation about the $\overline{z}$ axis. (Prime mark indicates derivative with respect to the spatial variable and the upper point indicates derivation with respect to time).

In addition, $E = E\left(\overline{x}\right)$ and $G = G\left(\overline{x}\right)$ are the Young's and shear modulus, while $\rho = \rho\left(\overline{x}\right)$ is the material density. The area and the second moment of area of the beam cross section are noted as $A = A\left(\overline{x}\right)$ and $I = I\left(\overline{x}\right)$. Lastly, the coefficient κ is the shear correction factor.

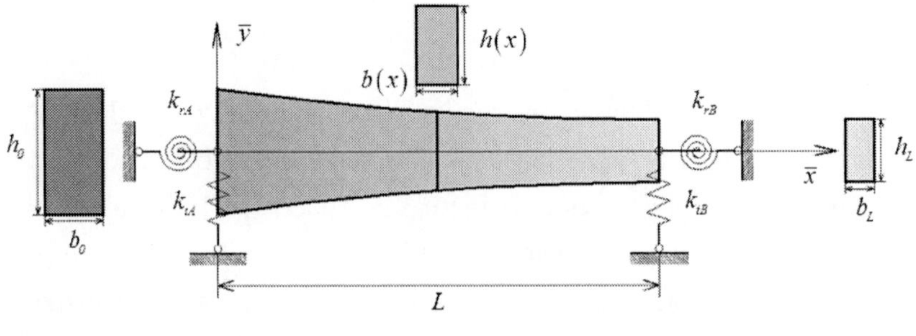

Figure 1. Vibrating AFG Timoshenko beam with varying cross section.

The boundary conditions for the Timoshenko beam model and taking into account the elastic restrains are given by:

$$\bar{x}=0: \quad \begin{array}{l} EI\dfrac{\partial \psi}{\partial x}=k_{rA}\,\psi \\ \kappa GA\left(\dfrac{\partial v}{\partial x}-\psi\right)=k_{tA}\,v \end{array} \qquad \bar{x}=L: \quad \begin{array}{l} EI\dfrac{\partial \psi}{\partial x}=-k_{rB}\,\psi \\ \kappa GA\left(\dfrac{\partial v}{\partial x}-\psi\right)=-k_{tB}\,v \end{array} \tag{5}$$

By assigning values to the stiffness coefficients of the translational and rotational springs, classical edge conditions can also be represented.

As the material properties and geometric characteristics of the beam may be general, one can define:

$$\begin{aligned} E(\bar{x})&=E_0\,f_E(\bar{x}) \\ G(\bar{x})&=G_0\,f_G(\bar{x}) \\ \rho(\bar{x})&=\rho_0\,f_\rho(\bar{x}) \\ I(\bar{x})&=I_0\,f_I(\bar{x}) \\ A(\bar{x})&=A_0\,f_A(\bar{x}) \\ b(\bar{x})&=b_0\,f_b(\bar{x}) \\ h(\bar{x})&=h_0\,f_h(\bar{x}) \end{aligned} \tag{6}$$

with: $f_A(\bar{x})=f_b(\bar{x})\times f_h(\bar{x})$, $f_I(\bar{x})=f_b(\bar{x})\times f_h^{\,3}(\bar{x})$ and $f_G(\bar{x})=f_E(\bar{x})$. It was considered:

$$G(\bar{x})=\frac{E(\bar{x})}{2(1+\mu)} \tag{7}$$

where μ is the Poisson ratio.

When the beam vibrates in one of its normal modes v and ψ can be written as:

$$v(\bar{x},t) = v = \bar{V}\, e^{i\omega t};\ \psi(\bar{x},t) = \psi = \bar{\Psi}\, e^{i\omega t} \tag{8}$$

where $\bar{V} = \bar{V}(\bar{x})$ and $\bar{\Psi} = \bar{\Psi}(\bar{x})$ are the spatial functions of the primary variables and ω is the circular natural frequency.

Replacing Eqs. (6) into Eqs. (2) and (3) the governing equations are obtained:

$$\left[\kappa AG\left(\bar{V}' - \bar{\Psi}\right)\right]' + \rho A\omega^2\bar{V} = 0 \tag{9}$$

$$\left(EI\bar{\Psi}'\right)' + \kappa AG\left(\bar{V}' - \bar{\Psi}\right) + \rho I\omega^2\bar{\Psi} = 0\,. \tag{10}$$

Non-dimensional expressions are introduced for spatial coordinate, displacement and rotation as follows:

$$x = \frac{\bar{x}}{L};\ V = \frac{\bar{V}(\bar{x})}{L};\ \Psi = \bar{\Psi}(\bar{x}); \tag{11}$$

and the natural non dimensional frequency coefficient is expressed as:

$$\Omega = \omega L^2\sqrt{\rho_0 A_0/(E_0 I_0)} \tag{12}$$

where $I_0 = I(0)$ and $A_0 = A(0)$.

Finally, the governing differential equations for an AFG Timoshenko vibrating beam become:

$$\frac{\kappa\, s^2}{2(1+\mu)}\left[\chi_1\left(V' - \Psi\right) + \chi_2\left(V'' - \Psi'\right)\right] + \Omega^2\, V = 0 \tag{13}$$

$$s^2\left[\chi_3\,\Psi' + \chi_2\,\Psi''\right] + \frac{s^4\,\kappa}{2(1+\mu)}\,\chi_4\,(V' - \Psi\,) + \Omega^2\,\Psi = 0 \tag{14}$$

where $s = L\sqrt{A_0/I_0}$ is the slenderness factor of the beam and $\chi_1, \chi_2, \chi_3, \chi_4$ are the variable coefficients, which can be written as follows:

$$\chi_1 = \frac{f_E{}'f_A + f_E\,f_A{}'}{f_\rho\,f_A};\ \chi_2 = \frac{f_E}{f_\rho};\ \chi_3 = \frac{f_E{}'f_I + f_E\,f_I{}'}{f_\rho\,f_I}\ ;\ \chi_4 = \frac{f_E\,f_A}{f_\rho\,f_I} \tag{15}$$

The Generalized Differential Quadrature Method

In order to obtain the GDQM analog equations of the governing equations of the AFG Timoshenko beam, the domain is discretized in a grid of N points using the Chebyshev–Gauss–Lobato expression (Zong and Zhang, 2009):

$$x_i = \frac{1}{2}\left\{1 - \cos\left[\frac{(i-1)\pi}{(N-1)}\right]\right\},\ i = 1, 2, \ldots, N\,; \tag{16}$$

where x_i is the coordinate of point i (Figure 2).

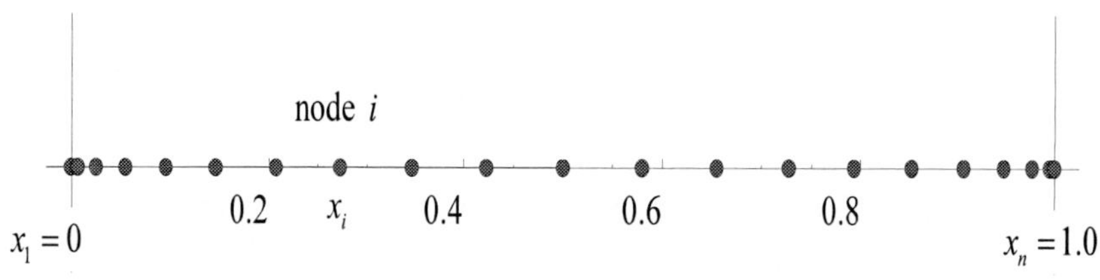

Figure 2. Grid of n points.

Based on the quadrature rules (Bert and Malik, 1997), the *qth* order derivatives of flexural displacement V and section rotation Ψ at a point i of the grid are expressed as:

$$\left.\frac{d^{(q)}V}{dx^q}\right|_{x_i} = \sum_{j=1}^{N} B_{ij}^{(q)} V_j ; \tag{17}$$

$$\left.\frac{d^{(q)}\Psi}{dx^q}\right|_{x_i} = \sum_{j=1}^{N} B_{ij}^{(q)} \Psi_j \tag{18}$$

where V_j and Ψ_j correspond to point j, and $B_{ij}^{(q)}$ are the weighting coefficients associated with the *qth* order derivative. They were obtained using Lagrange interpolating functions (Karami et al. 2003):

$$\Pi(x_i) = \prod_{j=1\ j\neq 1}^{N} \left(x_i - x_j\right) \tag{19}$$

$$B_{ij}^{(1)} = \frac{\Pi(x_i)}{(x_i - x_j)\,\Pi(x_j)},\ q=1;\ B_{ij}^{(q)} = B_{ii}^{(q-1)} B_{ij}^{(1)} - \frac{B_{ij}^{(q-1)}}{x_i - x_j},\ q>1;$$

with $i, j = 1, 2, ..., N,$ for $i \neq j$; (20)

$$B_{ii}^{(1)} = -\sum_{j=1\ j\neq i}^{N} B_{ij}^{(1)},\ q=1;\ B_{ii}^{(q)} = -\sum_{j=1\ j\neq i}^{N} B_{ij}^{(q)},\ q>1;$$

with $i, j = 1, 2, ..., N,$ for $i = j$. (21)

Using the quadrature rules, Eqs. (17) and (18), the differential quadrature analogs of the governing equations of a node i are:

$$-\frac{\kappa\, s^2}{2(1+\mu)}\left[\chi_{1i}\left(\sum_{j=1}^{N} B_{ij}^{(1)} V_j - \Psi_j\right) + \chi_{2i}\left(\sum_{j=1}^{N} B_{ij}^{(2)} V_j - \sum_{j=1}^{N} B_{ij}^{(1)} \Psi_j\right)\right] = \Omega^2 V_i \quad (22)$$

$$-s^2\left[\chi_{3i}\sum_{j=1}^{N} B_{ij}^{(1)}\Psi_j + \chi_{2i}\sum_{j=1}^{N} B_{ij}^{(2)}\Psi_j\right] - \frac{s^4\,\kappa}{2(1+\mu)}\chi_{4i}\left(\sum_{j=1}^{N} B_{ij}^{(1)} V_j - \Psi_j\right) = \Omega^2\,\Psi_i\,. \quad (23)$$

The preceding equations, together with the boundary conditions at the edges of the beam constitute the differential system that allows to determine the natural frequencies of the AFG Timoshenko beam, as well as its normal vibration modes.

RAYLEIGH-RITZ METHOD

If the cross section and constituent material vary smoothly, the energy functional J for the vibrating beam of length L is given by:

$$\begin{aligned} J(V,\Psi) = & \frac{1}{2}\left\{\int_0^1\left[\frac{E(x)I(x)}{L^2}\left(\frac{d\Psi(x)}{dx}\right)^2 + \kappa G(x)A(x)\left(\frac{dV(x)}{dx} - \Psi(x)\right)^2\right] L\,dx\right\} + \\ & + \frac{1}{2}\left\{k_{tA} L^2 \left(V\big|_{x=0}\right)^2 + k_{rA}\left(\Psi\big|_{x=0}\right)^2 + k_{tB} L^2\left(V\big|_{x=1}\right)^2 + k_{rB}\left(\Psi\big|_{x=1}\right)^2\right\} - \\ & - \frac{1}{2}\left\{\omega^2\int_0^1 \rho(x)\left[A(x)\left(V(x)\right)^2 L^2 + I(x)\left(\Psi(x)\right)^2\right] L\,dx\right\} \end{aligned} \quad (24)$$

To apply the Rayleigh-Ritz method, it is necessary to approximate the spatial component of the solution:

$$\begin{aligned} V(x) &\cong V_a(x) = \sum_{i=1}^{Np} C_i\, p_i(x) \\ \Psi(x) &\cong \Psi_a(x) = \sum_{j=1}^{Nq} D_j\, q_j(x) \end{aligned} \quad (25)$$

where $p_i(x)$ and $q_j(x)$ are coordinate functions that must satisfy at least the essential boundary conditions, C_i and D_j are arbitrary constants, Np and Nq are the number of terms.

Following Rayleigh-Ritz procedure, the functional is minimized with respect to every arbitrary constant (Ilanko and Monterrubio, 2014):

$$\begin{aligned} \frac{\partial J[V_a(x),\Psi_a(x)]}{\partial C_i} &= 0 \quad ; \quad i = 1,2,\ldots,Np. \\ \frac{\partial J[V_a(x),\Psi_a(x)]}{\partial D_j} &= 0 \quad ; \quad j = 1,2,\ldots,Nq. \end{aligned} \tag{26}$$

Then a linear system of equations is formed:

$$\mathbf{R}\begin{Bmatrix} C_i \\ D_j \end{Bmatrix} = \{0\} \tag{27}$$

which results in the following eigenvalue equation:

$$\mathbf{A} = \mathbf{K} - \Omega^2\,\mathbf{M} \tag{28}$$

with $\Omega = \omega L^2\sqrt{\rho_0 A_0 / E_0 I_0}$ the frequency coefficient and where:

$$\mathbf{K} = \begin{bmatrix} k_{ii} & k_{ij} \\ k_{ji} & k_{jj} \end{bmatrix} \tag{29}$$

with:

$$\begin{aligned} k_{ii} &= \frac{s^2\kappa}{2(1+\mu)}\int_0^1\left(f_E f_A\, p_i' p_i'\right)dx + K_{tA}\, p_i\big|_{x=0}\, p_i\big|_{x=0} + K_{tB}\, p_i\big|_{x=1}\, p_i\big|_{x=1} \\ k_{ij} &= -\frac{s^2\kappa}{2(1+\mu)}\int_0^1\left(f_E f_A\, p_i' q_j\right)dx \\ k_{ji} &= -\frac{s^2\kappa}{2(1+\mu)}\int_0^1\left(f_E f_A\, q_j\, p_i'\right)dx \\ k_{jj} &= \int_0^1\left(f_E f_I\, q_j' q_j'\right)dx + \frac{s^2\kappa}{2(1+\mu)}\int_0^1\left(f_E f_A\, q_j q_j\right)dx + K_{rA}\, q_j\big|_{x=0}\, q_j\big|_{x=0} + K_{rB}\, q_j\big|_{x=1}\, q_j\big|_{x=1} \end{aligned} \tag{30}$$

and

$$\mathbf{M} = \begin{bmatrix} m_{ii} & m_{ij} \\ m_{ji} & m_{jj} \end{bmatrix} \tag{31}$$

with

$$\begin{aligned} m_{ii} &= \int_0^1 \left(f_\rho f_A \, p_i \, p_i \right) dx \\ m_{ij} &= 0 \\ m_{ji} &= 0 \\ m_{jj} &= \frac{1}{s^2} \int_0^1 \left(f_\rho f_I \, q_j \, q_j \right) dx \end{aligned} \tag{32}$$

are the elements of matrices **K** and **M**, respectively; with

$$p_j' = \frac{dp_j}{dx}; q_j' = \frac{dq_j}{dx}; K_{tA} = \frac{k_{tA} L^3}{E_0 I_0}; K_{tB} = \frac{k_{tB} L^3}{E_0 I_0}; K_{rA} = \frac{k_{rA} L}{E_0 I_0}; K_{rB} = \frac{k_{rB} L}{E_0 I_0}. \tag{33}$$

Then, the eigenvalue problem can be expressed as:

$$\left| \mathbf{K}\mathbf{M}^{-1} - \Omega^2 \, \mathbf{I} \right| = \left| \mathbf{B} - \lambda \mathbf{I} \right| = 0 \tag{34}$$

where $\lambda = \Omega^2$ are the eigenvalues of matrix **B** and **I** the identity matrix.

The Finite Element Method

For this study a simple two-node element with four degrees of freedom is implemented, two in every node, (Reddy, 2002).

In the FEM, displacement and rotations fields of the element are associated with the nodal degrees of freedom by means of interpolation functions. The shape functions are obtained solving the differential equations of static equilibrium of a Timoshenko beam with uniform cross section and homogeneous material:

$$\kappa GA\left(\frac{d^2\bar{V}(\bar{x})}{d\bar{x}^2}-\frac{\partial\bar{\Psi}(\bar{x})}{d\bar{x}}\right)=0 \tag{35}$$

$$EI\frac{d^2\bar{\Psi}(\bar{x})}{d\bar{x}^2}+\kappa GA\left(\frac{d\bar{V}(\bar{x})}{d\bar{x}}-\bar{\Psi}(\bar{x})\right)=0. \tag{36}$$

Differentiating the Eq. (36) with respect to the spatial variable $\bar{x}$ and subtracting the derivative from Eq. (35):

$$\frac{d^3\bar{\Psi}(\bar{x})}{d\bar{x}^3}=0. \tag{37}$$

Then, differentiating twice the Eq. (35) with respect to the spatial variable $\bar{x}$ and replacing in Eq. (37), one obtains:

$$\frac{d^4\bar{V}(\bar{x})}{d\bar{x}^4}=0. \tag{38}$$

Substituting $\bar{x}=a\xi$, the solutions to Eqs. (37) and (38) are the complete polynomials:

$$V(\xi)=a_0+a_1\xi+a_2\xi^2+a_3\xi^3 \tag{39}$$

$$\Psi(\xi)=b_0+b_1\xi+b_2\xi^2 \tag{40}$$

which, introduced in Eqs. (35) and (36) yields to

$$\frac{1}{a}(2a_2 + 6a_3\,\xi) - \frac{1}{a}(b_1 + 2b_2\,\xi) = 0 \tag{41}$$

$$\frac{\kappa GA}{a}(a_1 + 2a_2\,\xi + 3a_3\,\xi^2) - \frac{EI}{a^2}(2b_2) - \kappa GA(b_0 + b_1\xi + b_2\,\xi^2) = 0\,. \tag{42}$$

Eqs. (41) and (42) must be satisfied for every point of the element, then:

$$b_0 = \frac{1}{a}a_1 + \frac{2\phi}{a}a_3 \;\; ; \;\; b_1 = \frac{2}{a}a_2 \;\; ; \;\; b_2 = \frac{3}{a}a_3 \quad \text{con} \quad \phi = \frac{3EI}{\kappa GAa^2}\,. \tag{43}$$

Taking derivative of Eq. (39) with respect to ξ and substituting Eq. (43) into Eq. (40), the expression of distortion γ is:

$$\gamma(\xi) = \frac{1}{a}\frac{\partial V}{\partial \xi} - \Psi = -\frac{2\phi\,a_3}{a}\,. \tag{44}$$

From Eq. (44) it can be seen that the distortion is constant along the element, while the shear deformation varies linearly.

Then, the four degree of freedom of the element are of the form:

$$\begin{aligned} V_1 &= V(-1) = a_0 - a_1 + a_2 - a_3 \\ V_2 &= V(+1) = a_0 + a_1 + a_2 + a_3 \\ \Psi_1 &= V(-1) = \left(a_1 + 2\phi\,a_3 - 2\,a_2 + 3\,a_3\right)/\,a \\ \Psi_2 &= V(+1) = \left(a_1 + 2\phi\,a_3 + 2\,a_2 + 3\,a_3\right)/\,a\;. \end{aligned} \tag{45}$$

Solving the linear system of equations coefficients a_0, a_1, a_2, a_3 are determined as functions of the degree of freedom V_1, Ψ_1, V_2 and Ψ_2. Then they are introduced in Eqs. (39) and (40) and lead to:

$$V(\xi)=\left[N_{v1}(\xi)\ N_{v2}(\xi)\ N_{v3}(\xi)\ N_{v4}(\xi)\right]\begin{Bmatrix}V_1\\ \Psi_1\\ V_2\\ \Psi_2\end{Bmatrix}=\left[N_v(\xi)\right]\{\delta_e\} \tag{46}$$

$$\Psi(\xi)=\left[N_{\psi1}(\xi)\ N_{\psi2}(\xi)\ N_{\psi3}(\xi)\ N_{\psi4}(\xi)\right]\begin{Bmatrix}V_1\\ \Psi_1\\ V_2\\ \Psi_2\end{Bmatrix}=\left[N_\psi(\xi)\right]\{\delta_e\} \tag{47}$$

where the components of $\{\delta_e\}$ are the degree of freedom (D.O.F) of the element and the eight interpolation functions are:

$$\begin{aligned}
N_{v1}(\xi)&=\frac{1}{4(1+\phi)}\left[2(1+\phi)-3(1+2\phi)\xi+\xi^3\right]\\
N_{v2}(\xi)&=\frac{a}{4(1+\phi)}\left[1+\phi-\xi-(1+\phi)\xi^2+\xi^3\right]\\
N_{v3}(\xi)&=\frac{1}{4(1+\phi)}\left[2(1+\phi)+3(1+2\phi)\xi-\xi^3\right]\\
N_{v4}(\xi)&=\frac{a}{4(1+\phi)}\left[-1-\phi-\xi+(1+\phi)\xi^2+\xi^3\right]
\end{aligned} \tag{48}$$

$$\begin{aligned}
N_{\psi1}(\xi)&=\frac{3}{4(1+\phi)}\left[-1+\xi^2\right]\\
N_{\psi2}(\xi)&=\frac{1}{4(1+\phi)}\left[-1+2\phi-2(1+\phi)\xi+3\xi^2\right]\\
N_{\psi3}(\xi)&=\frac{3}{4(1+\phi)}\left[1-\xi^2\right]=-N_{\psi1}(\xi)\\
N_{\psi4}(\xi)&=\frac{1}{4(1+\phi)}\left[-1+2\phi+2(1+\phi)\xi+3\xi^2\right]
\end{aligned} \tag{49}$$

for $\xi=\bar{x}/a,(-1\le\xi\le1)$. The distortion γ of the element is expressed:

$$\gamma(\xi)=\frac{1}{a}\frac{\partial V(\xi)}{\partial \xi}-\Psi(\xi)=\left[N_{\gamma 1}(\xi)\ N_{\gamma 2}(\xi)\ N_{\gamma 3}(\xi)\ N_{\gamma 4}(\xi)\right]\begin{Bmatrix} V_1 \\ \Psi_1 \\ V_2 \\ \Psi_2 \end{Bmatrix}=\frac{1}{a}\left[N_\gamma(\xi)\right]\{\delta_e\} \tag{50}$$

with $N_{\gamma 1}=-N_{\gamma 3}=-\frac{\phi}{2(1+\phi)a}$; $N_{\gamma 2}=N_{\gamma 4}=a\,N_{\gamma 1}$.

Then, the mass matrix $\left[m_e\right]$ and the stiffness matrix $\left[k_e\right]$ of the element are obtained.

To obtain the mass matrix the expression of the kinetic energy is considered:

$$T_e=\frac{1}{2}\omega^2\left\{\int_{-1}^{1}\rho(\xi)A(\xi)\left(V(\xi)\right)^2 a\,d\xi+\int_{-1}^{1}\rho(\xi)I(\xi)\left(\Psi(\xi)\right)^2 a\,d\xi\right\}. \tag{51}$$

Expressing the displacements in terms of vectors $\left[N_w(\xi)\right]$ and $\left[N_\psi(\xi)\right]$:

$$\left(V(\xi)\right)^2=\{\delta_e\}^t\left[N_v(\xi)\right]^t\left[N_v(\xi)\right]\{\delta_e\}\,;\left(\Psi(\xi)\right)^2=\{\delta_e\}^t\left[N_\psi(\xi)\right]^t\left[N_\psi(\xi)\right]\{\delta_e\} \tag{52}$$

And substituting in Eq. (51) is obtained:

$$\begin{aligned} T_e &=\frac{1}{2}\omega^2\left(\{\delta_e\}^t\int_{-1}^{1}\rho(\xi)A(\xi)\left[N_v(\xi)\right]^t\left[N_v(\xi)\right]a\,d\xi\,\{\delta_e\}\right)+ \\ &+\frac{1}{2}\omega^2\left(\{\delta_e\}^t\int_{-1}^{1}\rho(\xi)I(\xi)\left[N_\psi(\xi)\right]^t\left[N_\psi(\xi)\right]a\,d\xi\{\delta_e\}\right) \\ &=\frac{1}{2}\omega^2\{\delta_e\}^t\left[m_e\right]\{\delta_e\} \end{aligned} \tag{53}$$

with

$$\left[m_e\right]=a\left(\int_{-1}^{1}\rho(\xi)A(\xi)\left[N_w(\xi)\right]^t\left[N_w(\xi)\right]d\xi+\int_{-1}^{1}\rho(\xi)I(\xi)\left[N_\psi(\xi)\right]^t\left[N_\psi(\xi)\right]d\xi\right).$$

To obtain the stiffness matrix the expression of the potential energy must be considered:

$$U_e = \frac{1}{2}\left\{ \int_{-1}^{1} E(\xi) I(\xi) \left(\frac{1}{a} \Psi(\xi)' \right)^2 a\, d\xi + \kappa \int_{-1}^{1} G(\xi) A(\xi) \left(\frac{1}{a} V(\xi)' - \Psi(\xi) \right)^2 a\, d\xi \right\} \tag{54}$$

the factors that contain the displacements are expressed taking into account vectors $\left[N_\psi'(\xi) \right]$ and $\left[N_\gamma(\xi) \right]$:

$$\begin{aligned} \left(\frac{1}{a} \Psi(\xi)' \right)^2 &= \frac{1}{a^2} \{\delta_e\}^t \left[N_\psi'(\xi) \right]^t \left[N_\psi'(\xi) \right] \{\delta_e\} \\ \left(\frac{1}{a} V(\xi)' - \Psi(\xi) \right)^2 &= (\gamma(\xi))^2 = \frac{1}{a^2} \{\delta_e\}^t \left[N_\gamma(\xi) \right]^t \left[N_\gamma(\xi) \right] \{\delta_e\} \end{aligned} \tag{55}$$

and replacing in Eq. (54) results:

$$\begin{aligned} U_e &= \frac{1}{2}\left(\{\delta_e\}^t \int_{-1}^{1} \frac{E(\xi) I(\xi)}{a^2} \left[N_\psi'(\xi) \right]^t \left[N_\psi'(\xi) \right] a\, d\xi \{\delta_e\} \right) + \\ &+ \frac{1}{2}\left(\{\delta_e\}^t \kappa \int_{-1}^{1} \frac{E(\xi) I(\xi)}{a^2} \left[N_\psi'(\xi) \right]^t \left[N_\psi'(\xi) \right] a\, d\xi \{\delta_e\} \right) \\ &= \frac{1}{2} \{\delta_e\}^t [k_e] \{\delta_e\} \end{aligned} \tag{56}$$

with

$$[k_e] = \frac{1}{a}\left(\int_{-1}^{1} E(\xi) I(\xi) \left[N_\psi'(\xi) \right]^t \left[N_\psi'(\xi) \right] d\xi + \kappa \int_{-1}^{1} G(\xi) A(\xi) \left[N_\gamma(\xi) \right]^t \left[N_\gamma(\xi) \right] d\xi \right).$$

By assembling the matrices of the elements $[m_e]$ and $[k_e]$ for the entire domain of the beam, the matrices **M** and **K** are obtained. With them, the characteristic equation that provides the values of the natural frequencies is constructed:

$$\{\mathbf{K} - \omega^2 \mathbf{M}\} \{\delta_e\} = \{0\} . \tag{57}$$

VERIFICATION OF THE PROPOSED MODEL

In order to verify the accuracy of the proposed model, comparisons is made with particular cases available in the literature for ideal boundary conditions. First, Table 1 compares values for a tapered Timoshenko beam made of axially functionally graded material studied by Shahba et al. (2011), Huang et al. (2013), and Zhao et al. (2017). They obtained values for the first four natural frequency coefficients for a case that can be represented in the present model. The two constituent materials are Zirconia (ZrO_2) and Aluminum (Al).

$$ZrO_2\text{:}\ E_a = 200\ \text{GPa}\ ;\ \rho_a = 5700\ \frac{kg}{m^3} \quad (58)$$

$$\text{Al:}\ E_b = 2702\,\text{GPa}\ ;\ \rho_b = 2702\frac{kg}{m^3} \quad (59)$$

and properties of AFG material, like mass density ρ, Young's modulus E, shear modulus G, continuously vary in the axial direction with a power law relation. In order to compare results, it is taken $n = 2$ in Eq. (2). In this case: $E_0 = E_a$. The geometry of the varying section is defined adopting in Eqs. (6):

$$f_A = 1 - 0.1x\ , f_I = (1 - 0.1x)^3\ . \quad (60)$$

They adopted: $s = 10; \kappa = 5/6; \mu = 0.30$

In order to represent the known ideal supports, the following values are adopted for the elastic springs in Figure 1:

- Clamped: $k_t \to \infty\ ,\ k_r \to \infty$
- Pinned: $k_t \to \infty\ ,\ k_r = 0$
- Free: $k_t = 0\ ,\ k_r = 0$
- Guided: $k_t = 0\ ,\ k_r \to \infty$

Table 1. Frequency coefficients $\Omega_i = \omega_i L^2 \sqrt{(\rho_0 A_0)/(E_0 I_0)}$ for a tapered AFG Timoshenko beam

Boundary Conditions	Ω	GDQM	R-Ritz M	FEM	Shahba et al. (2011)	Huang et al. (2013)	Zhao et al. (2017)
Clamped-Free	1	3.93579	3.93579	3.93574	3.9359	3.93579	3.93585
	2	15.1533	15.1533	15.1533	15.1577	15.1533	15.1540
	3	31.2239	31.2239	31.2239	31.2638	31.2239	31.2257
	4	47.5836	47.5836	47.5837	47.7164	47.5857	47.5871
	5	62.7344	62.7344	62.7346	–	–	–
Clamped-Pinned	1	10.8007	10.8007	10.8007	–	10.8007	10.8011
	2	25.6179	25.6179	25.6180	–	25.6179	25.6194
	3	42.6474	42.6474	42.6476	–	42.6478	42.6502
	4	58.8528	58.8528	58.8532	–	58.8595	58.8588
	5	62.7800	62.7800	62.7802	–	–	–
Clamped-Clamped	1	12.4633	12.4633	12.4634	12.4689	12.4633	12.4691
	2	26.3804	26.3804	26.3806	26.4153	26.3804	26.3822
	3	42.9607	42.9607	42.9609	43.0904	42.9611	42.9636
	4	59.3916	59.3916	59.3919	59.6829	59.4023	59.3958
	5	68.0580	68.0580	68.0583	–	–	–
Pinned-Pinned	1	7.65276	7.65276	7.65281	7.6545	–	7.6529
	2	23.7126	23.7126	23.7127	23.7369	–	–
	3	41.7042	41.7042	41.7044	41.8210	–	–
	4	57.7613	57.7613	57.7616	57.8739	–	–
	5	60.1514	60.1514	60.1517	–	–	–

A very good correlation is observed between the results. The results were obtained using 23 nodes for GDQM, 6000 elements for FEM and in the case of Rayleigh-Ritz it was taken in Eq. (24) N_p=N_q=20 and complete polynomials of the second degree:

$$p_i(x) = \left(\alpha_1 x^2 + \alpha_2 x + 1\right) x^{i-1}$$
$$q_j(x) = \left(\beta_1 x^2 + \beta_2 x + 1\right) x^{j-1} \quad (61)$$

For boundary conditions that lead to unstable beams, no results have been found for AFG Timoshenko beams. Therefore, comparison is made with the results of the homogeneous Bernoulli-Euler beams with constant cross section available in the thorough Karnovsky-Lebed (2004) reference manual.

Table 2. Eigenvalues $\eta = \sqrt{\Omega}$ for unstable Bernoulli-Euler beams

Boundary Condition	η	GDQM	R-Ritz M	FEM	Karnovsky Lebed (2004)
Free-Free	1	4.73002	4.73002	4.73002	4.73004
	2	7.85313	7.85313	7.85313	7.85320
	3	10.9954	10.9954	10.9954	10.9956
	4	14.1367	14.1367	14.1367	14.1372
Free-Pinned	1	3.92659	3.92659	3.92659	3.92660
	2	7.06852	7.06852	7.06852	7.06858
	3	10.2100	10.2100	10.2100	10.2102
	4	13.3514	13.3514	13.3514	13.3518
Free-Guided	1	2.36502	2.36502	2.36502	2.36502
	2	5.49778	5.49778	5.49778	5.49780
	3	8.63927	8.63927	8.63927	8.63938
	4	11.7807	11.7807	11.7807	11.7810
Guided-Guided	1	3.14159	3.14159	3.14159	3.14159
	2	6.28314	6.28314	6.28314	6.28319
	3	9.42464	9.42464	9.42464	9.42478
	4	12.5660	12.5660	12.5661	12.5664

The comparison is useful to verify that the proposed model can also satisfactorily represent a Bernoulli-Euler beam. For this, it is necessary to assign a suitable large slenderness value. It was found that for $s = 2500$ the results obtained are in good agreement with those of the Bernoulli-Euler beam. Table 2 shows the first four non-zero eigenvalues for the indicated unstable beam cases. In all cases, the null values corresponding to the movement of the rigid body have been omitted. As can be seen, the obtained values are adequately close to those of the Bernoulli-Euler beam for all implemented solutions.

NUMERICAL RESULTS

The quantity and variability of the parameters involved in the description of the mechanical behavior of these kinds of materials as well as in the representation of beam geometries and in their external supports make of little interest the construction of tables of dynamical magnitudes, as it is usual for homogeneous beams with ideal boundary conditions. Consequently, just a few representative cases will be considered to demonstrate the convenience of the procedure.

All calculations are performed using 23 nodes for GDQM, 6000 elements for FEM and in the case of Rayleigh-Ritz it was taken in Eq. (24) $N_p = N_q = 20$.

Unstable AFG Timoshenko Beams

As no previous results have been found for the case of unstable AFG beams, a particular case is analized. For the same functionally graded material of Table 1, a combination of Zirconia and Aluminum that varies axially according to an exponential law, frequency coefficients are obtained for the unstable beams of Table 2. The results, excluding the null values, are shown in Table 3:

Table 3. Frequency coefficients $\Omega_i = \omega_i L^2 \sqrt{(\rho_0 A_0)/(E_0 I_0)}$ for unstable AFG Timoshenko beam (s=10, κ=5/6, μ=0.3)

Boundary Condition	Ω_1	Ω_2	Ω_3	Ω_4	Ω_5	Solution
Free-Free	16.5511	33.3001	50.6019	59.6066	72.6755	GDQM
	16.5511	33.3001	50.6019	59.6066	72.6755	R-Ritz M
	16.5516	33.3009	50.6032	59.6074	72.6772	FEM
Pinned-Free	12.6635	29.5292	47.2029	64.3574	72.6815	GDQM
	12.6635	29.5292	47.2029	64.3574	72.6815	R-Ritz M
	12.6638	29.5300	47.2041	64.3586	72.6832	FEM
Guided-Free	5.76045	20.8414	38.4271	55.4807	64.8920	GDQM
	5.76045	20.8414	38.4271	55.4807	64.8920	R-Ritz M
	5.76063	20.8420	38.4281	55.4820	64.8931	FEM
Guided-Guided	7.94687	24.1429	42.2119	60.5274	67.1745	GDQM
	7.94687	24.1429	42.2119	60.5274	67.1745	R-Ritz M
	7.94717	24.1437	42.2134	60.5295	67.1749	FEM

It is worth mentioning that the Rayleigh-Ritz method gives upper bounds of the wanted values. All three approaches show agreement on numerical values.

Simply Supported Beam

Variation in the Cross Section

As is known, the variation of the cross section of a beam introduces modifications in the values of its natural frequencies. Symmetrical variations in beam height are considered according to a second-degree polynomial law.

$$h(x) = h_0 + 4(h_2 - h_0)(1 - x)x \tag{62}$$

where h_0 is the beam hcight at the edges and h_2 is the beam height at the middle section.

Three cases are evaluated for an homogeneous simply supported (pinned-pinned) Timoshenko beam with $s = 100$; $\kappa = 5/6$ and $\mu = 0.30$.

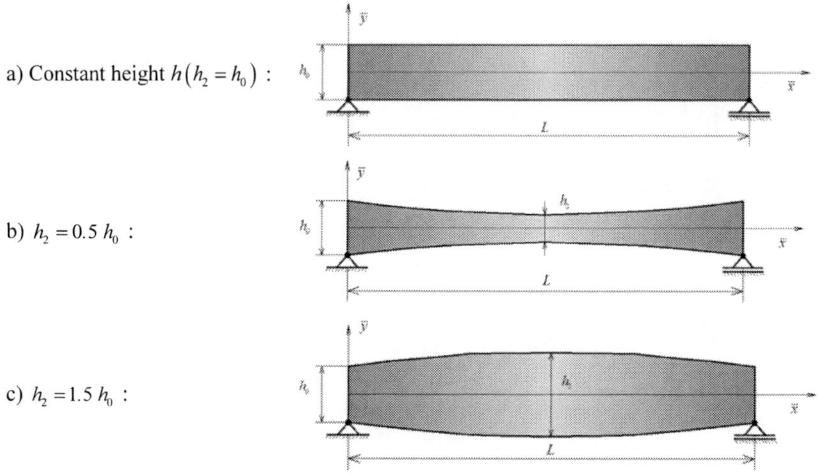

Figure 3. Geometries considered for simply supported beams.

Table 4. Frequency coefficients $\Omega_i = \omega_i L^2 \sqrt{(\rho_0 A_0)/(E_0 I_0)}$ for a simply supported homogeneous Timoshenko beam

Case	h_2/h_0	Ω1	Ω2	Ω3	Ω4	Ω5	Solution
a)	1	9.84961	39.1621	87.2544	153.067	235.265	GDQM
		9.84961	39.1621	87.2544	153.067	235.264	R-Ritz M
		9.84961	39.1621	87.2544	153.067	235.265	FEM
b)	0.5	5.36446	24.0757	56.1379	100.031	155.520	GDQM
		5.36446	24.0757	56.1379	100.031	155.520	R-Ritz M
		5.36446	24.0757	56.1375	100.031	155.520	FEM
c)	1.5	13.9653	52.1650	114.748	198.883	301.982	GDQM
		13.9653	52.1650	114.748	198.883	301.982	R-Ritz M
		13.9653	52.1651	114.748	198.883	301.982	FEM

The results shown in Table 4.

Taking the uniform beam a) as a reference for case c) the frequencies increase, the first by 42% and on average 33% and for b) they decrease on average by 40% and the first one by 45%.

In this regard, it is necessary to point out that if we call $W_0 = \rho_0\, g\, b_0\, h_0\, L$ to the weight of the beam a), and considering the same value for h_0 in the other two cases, it will be $W_{b)} = \frac{2}{3} W_0$ and $W_{c)} = \frac{4}{3} W_0$. Therefore the increase or decrease in the frequency values is of the same order as the increase or decrease in weight.

Certainly, if we assign in b) and c) values to h_0 and h_2 in order to maintain the established relations h_2/h_0 = 0.5, 1.5 and the same magnitude of the total weight of the uniform beam $W_0 = \rho_0\, g\, b_0\, h_0\, L$, we will have:

- Case a): $h(0) = 1.5\, h_0,\ h\left(L/2\right) = 0.75\, h_0$
- Case b): $h(0) = 0.75\, h_0,\ h\left(L/2\right) = 1.125\, h_0$

and the frequency coefficient values are shown in Table 5:

Table 5. Frequency coefficients $\Omega_i = \omega_i L^2 \sqrt{(\rho_0 A_0)/(E_0 I_0)}$ for a simply supported homogeneous Timoshenko beam with varying cross section and weight $W_0 = \rho_0\, g\, b_0\, h_0\, L$

Case	Ω_1	Ω_2	Ω_3	Ω_4	Ω_5	Solution
b)	8.04669	36.1136	84.2069	150.047	233.280	GDQM
	8.04669	36.1136	84.2069	150.047	233.280	R-Ritz M
	8.04669	36.1136	84.2063	150.047	233.280	FEM
c)	10.4740	39.1238	86.0610	149.162	226.487	GDQM
	10.4740	39.1238	86.0610	149.162	226.487	R-Ritz M
	10.4740	39.1238	86.0610	149.162	226.487	FEM

It can be observed in table 5 when considering case c) without growing the weight of the beam, that just the first frequency increases (and only 6%), the second one remains practically the same and the rest decrease.

Therefore, changing the shape of a beam does not constitute an effective way to stiffen it since it is also required to increase its weight which can be disandvantageous for the rest of the structure.

Variation in the Material

The use of functional materials provides an excellent solution to this problem. Consider than the beam is built with the AFG material made of steel and aluminum oxide Al_2O_3 (alumina) proposed by Su et al. (2013).

Their Young modulus and density are:

$$St:\ E_{St} = 210 \text{ GPa}\,;\ \rho_{St} = 7800\ \frac{kg}{m^3}\,;\ \nu_{St} = 0.30 \tag{63}$$

$$Alum:\ E_{Alum} = 390 \text{ GPa}\,;\ \rho_{Alum} = 3960\frac{kg}{m^3}\,;\ \nu_{Alum} = 0.30 \tag{64}$$

The relationships between material properties are: $E_{Alum}/E_{St} = 1.857$ for Young's modulus and $\rho_{Alum}/\rho_{St} = 0.508$ for the density. Note that the alumina, more rigid, is lighter than steel.

In order to evaluate the influence of the material composition, an AFG beam of uniform cross section is analyzed.

The inhomogeneous material, with gradient compositional variation of the constituents (Steel and Aluminna) varies in the longitudinal direction of the beam. Properties of AFG materials, like mass density ρ, Young's modulus E, shear modulus G, continuously vary in the axial direction according to Eq. (1), therefore, there will be entire sections of steel at the ends and full section of alumina in the center of the beam.

In all evaluated cases, in order to facilitate comparisons will be considered $E_0 = E_{St}$ and $\rho_0 = \rho_{St}$.

The weight of the beam is considered in relation to the uniform beam of steel, which is taken as reference material:

$$W_b = \frac{g\int_0^L \rho\, b\, h\, d\bar{x}}{\rho_{st} g b_0 h_0 L} \tag{65}$$

Comparing with Table 4, case a), the increase is remarkable, of 74% in the first frequency and above 50% in the rest. It is also interesting to note that the weight of this AFG beam is $W_b = 0.67$, that is 33% lighter than the uniform steel beam.

In this case the combination is doubly beneficial because in the central portion of the beam where the curvatures and displacements are greater, the majority presence of alumina increases the potential energy due to its greater rigidity and decreases the kinetic energy due to its lower weight. That makes natural frequencies increase.

Table 6. Frequency coefficients $\Omega_i = \omega_i L^2 \sqrt{(\rho_0 A_0)/(E_0 I_0)}$ for a simply supported AFG uniform Timoshenko beam (s=100, κ=5/6, μ=0.3)

Ω_1	Ω_2	Ω_3	Ω_4	Ω_5	Solution
17.1099	61.1944	134.383	234.256	358.928	GDQM
17.1099	61.1944	134.383	234.256	358.930	R-Ritz M
17.1099	61.1944	134.383	234.256	358.928	FEM

Table 7. Frequency coefficients $\Omega_i = \omega_i L^2 \sqrt{(\rho_0 A_0)/(E_0 I_0)}$ for a simply supported AFG Timoshenko beam of varying cross section

Case	h_2/h_0	Ω_1	Ω_2	Ω_3	Ω_4	Ω_5	Solution
b)	0.5	9.39781	38.4923	88.8831	156.942	242.782	GDQM
		9.39781	38.4923	88.8831	156.942	242.782	R-Ritz M
		9.39781	38.4923	88.8831	156.942	242.782	FEM
c)	1.5	24.0553	80.2770	174.352	300.670	455.568	GDQM
		24.0553	80.2770	174.352	300.670	455.568	R-Ritz M
		24.0553	80.2770	174.352	300.670	455.568	FEM

Variation in Section and Material

Then, the effect caused by varying both the section and the material composition is analyzed, cases b) and c) of Figure 3 and the material composition of the last case.

The results are shown in Table 7.

As can be seen, the results for case b) are rather similar to those of the steel beam of uniform section (Table 4). Although it is important to note that the weight of the beam in this case is $W_b = 0.45$, i.e., similar natural frequencies with less than half the weight.

For the case c) the fundamental frequency increases by 144% and the rest doubles those of the uniform steel beam. The weight of the beam is $W_b = 0.90$.

Cantilever Beam

Finally, the case of a cantilver beam is studied (Figure 4). We analyze a case with a cross section varying both in width and height and with an AFG material combination of Alumina and Steel. All the variances are assumed according to Eq. (2) with linear law ($n = 1$).

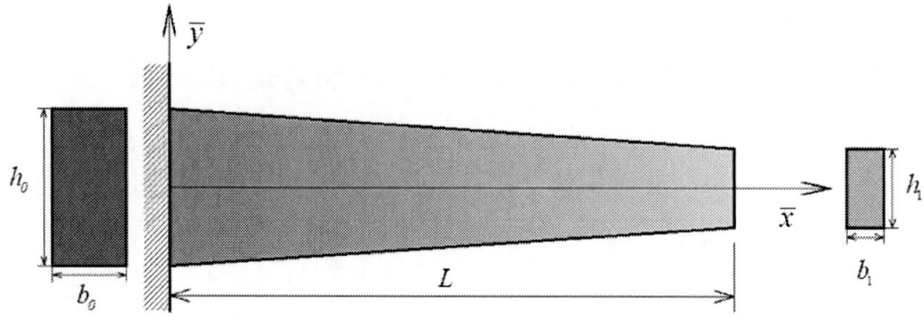

Figure 4. Cantilever AFG Timoshenko beam.

Thus, for $\bar{x}=0$ the entire section is of Alumina and for $\bar{x}=L$ the whole section is of Steel. And a typical material property $P(\bar{x})$ such as ρ, E or G assumes the variation:

$$P(\bar{x}) = P_{Alum} + (P_{St} - P_{Alum})\left(\frac{\bar{x}}{L}\right) \tag{66}$$

To facilitate comparison, the frequency coefficients refer to the properties of steel $\Omega_i = \omega_i L^2 \sqrt{\dfrac{\rho_{St} A_0}{E_{St} I_0}}$, it was taken $s = 100$; $\kappa = 5/6$ and $\mu = 0.30$.

Values have been found for the first two frequency coefficients and are indicated in Table 8. The reference values for the homogeneous uniform cantilever Timoshenko beam are (Karnovsky-Lebed, 2004):

Ω_1=3.51265
Ω_2=21.8889

Analyzing the values, as a first conclusion it must be said that all the frequency coefficients are increased with respect to the uniform steel cantilever beam.

Table 8. Frequency coefficients for an AFG Timoshenko cantilever beam

b_1/b_0	1.0		0.8		0.6		0.4		0.2		Solution
h_1/h_0	Ω_1	Ω_2	Ω_1	Ω_2	Ω_1	Ω_2	Ω_1	Ω_2	Ω_1	Ω_2	
1.0	4.77959	29.8245	4.88453	27.7697	5.03312	25.5650	5.26541	23.1621	5.70360	20.5136	GDQM
	4.77959	29.8245	4.88453	27.7697	5.03312	25.5650	5.26541	23.1621	5.70360	20.5136	R-Ritz M
	4.77943	29.8233	4.88437	27.7686	5.03295	25.5639	5.26521	23.1611	5.70339	20.5128	FEM
0.8	5.11525	30.4578	5.21793	28.3574	5.36403	26.1049	5.59385	23.6511	6.03106	20.9455	GDQM
	5.11525	30.4578	5.21793	28.3574	5.36403	26.1049	5.59385	23.6511	6.03106	20.9455	R-Ritz M
	5.11510	30.4566	5.21772	28.3563	5.36383	26.1039	5.59360	23.6501	6.03086	20.9447	FEM
0.6	5.57304	31.2932	5.67290	29.1334	5.81582	26.8188	6.04225	24.2986	6.47720	21.5194	GDQM
	5.57304	31.2932	5.67290	29.1334	5.81582	26.8188	6.04225	24.2986	6.47720	21.5194	R-Ritz M
	5.57280	31.2919	5.67269	29.1322	5.81562	26.8178	6.04204	24.2976	6.47699	21.5186	FEM
0.4	6.25055	32.5138	6.34709	30.2690	6.48614	27.8651	6.70814	25.2496	7.13867	22.3658	GDQM
	6.25055	32.5138	6.34709	30.2690	6.48614	27.8651	6.70814	25.2496	7.13867	22.3659	R-Ritz M
	6.25033	32.5125	6.34688	30.2677	6.48591	27.8639	6.70790	25.2436	7.13844	22.3650	FEM
0.2	7.40467	34.7340	7.49886	32.3423	7.63496	29.7832	7.85282	27.0017	8.27545	23.9379	GDQM
	7.40467	34.7340	7.49886	32.3423	7.63496	29.7832	7.85282	27.0017	8.27545	23.9380	R-Ritz M
	7.40439	34.7326	7.49855	32.3410	7.63467	29.7820	7.85251	27.0006	8.27516	23.9370	FEM

The case $b_1/b_0 = h_1/h_0 = 1$ that corresponds to an AFG beam of uniform cross section shows the effect of the FG material.

The rest of the values combine the effects of the beam geometry and the composition of the material.

The effect of the geometry clearly emerges: As the height decreases away from the clamped edge, the first and second frequencies grow. On the other hand, when the width is reduced, the first frequency grows but the second frequency decreases.

The growth of the first frequency for both cases is explained taking into account the shape of the first mode of vibration (without nodes): the inertial mass is reduced in the zones of greater displacement and consequently the kinetic energy is reduced.

As for the second frequency, the existence of nodes in the second mode of vibration causes significant curvatures far from the clamped edge and the effect of reducing the height on the potential energy is greater than the effect of the reduction of the width.

Again, the agreement between the three approaches is excellent.

In Table 9 are indicated the weight of every beam considered in Table 8 in relation to the uniform beam of steel, which is taken as reference material -Eq. (65)-:

Table 9. Weight of the AFG beam – Eq. (65)-

		b_1/b_0				
		1.0	0.8	0.6	0.4	0.2
h_1/h_0	1.0	0.75	0.67	0.59	0.50	0.42
	0.8	0.67	0.60	0.53	0.45	0.38
	0.6	0.59	0.53	0.47	0.41	0.35
	0.4	0.50	0.45	0.41	0.36	0.31
	0.2	0.42	0.38	0.35	0.31	0.27

It is possible to observe that the use of FG materials allows a remarkable increase in the dynamic stiffness of the beam with an amazing decrease in its weight.

CONCLUSION

As it has been seen, the functionally graded materials (FGM) used in resistant structures allow to develop excellent performances in vibratory environments, which has enormous interest in numerous technological applications in various fields of engineering (civil, mechanical, electronic, etc.) when looking for simultaneously rigidity and lightness.

The proposed model through its generality can represent innumerable cases of vibrating beams: variations in geometry, different combinations of different constituent materials and all the possibilities of restraints at its edges (including the classic ideals boundary conditions). As known, ideal supports used in many structural models do not fit exactly the real bonding of the beam with the rest of the structure which are, generally, elastic.

Three classical approaches, the variational method of Rayleigh-Ritz, the Differential Quadrature method and a development in Finite Elements implemented by the authors have been successfully used in the present study to obtain approximate, yet quite accurate, solutions to a difficult elastodynamics problem with modern technological applications. The Rayleigh-Ritz and GDQM approaches have shown a more accurate precision in the frequency values obtained with a lower computational cost, than the finite element approach implemented.

ACKNOWLEDGMENTS

The present study has been sponsored by the Universidad Nacional del Sur, UNS, the Consejo Nacional de Investigaciones Científicas y Técnicas,

CONICET and the Comisión de Investigaciones Científicas, Buenos Aires Province (CIC). It was developed in the Department of Engineering.

REFERENCES

Alshorbgy, A. E., Eltaher, M. A. and Mahmoud, F. F. (2011). Free vibration characteristics of a functionally graded beam by finite element method. *Applied Mathematical Modelling 35*, 412–425.

Aydogdu, M. and Taskin, V. (2007). Free vibration analysis of functionally graded beams with simply supported edges. *Materials and Design 28*, 1651–1656.

Bambill D. V., Rossit C. A. and Felix D. H. (2015). Free vibrations of stepped axially functionally graded Timoshenko beams. *Meccanica, 50*, 1073–1087.

Banerjee J. R. (2001). Dynamic stiffness formulation and free vibration analysis of centrifugally stiffened Timoshenko beam. *Journal of Sound and Vibration, 247*, 97-115.

Bert, C. W. and Malik, M. (1997). Differential quadrature: a powerful new technique for analysis of composite structures. *Composite Structures 39(3–4)*, 179–189.

Calio, I. and Elishakoff, I. (2004). Closed-form trigonometric solutions for inhomogeneous beam-columnson elastic foundation. *International Journal of Structural Stability and Dynamics 4*, 139–146.

Calio I. and Elishakoff I. (2005). Closed-form solutions for axially graded beam-columns. *Journal of Sound and Vibration 280*, 1083–1094.

Çetin, D. and Simsek, M. (2011). Free vibration of an axially functionally graded pile with pinned ends embedded in Winkler-Pasternak elastic medium. *Structural Engineering and Mechanics 40*, 583–594.

Chabraborty, A., Gopalakrishnan, S. and Reddy, J. N. (2003). A new beam finite element for the analysis of functionally graded materials. *International Journal of Mechanical Sciences 45*, 519–539.

Chegenizadeh A., Ghadimi B., Nikraz H. and Şimşek M. (2014). A novel two-dimensional approach to modelling functionally graded beams resting on a soil medium. *Structural Engineering and Mechanics, 51(5)*, 727–741.

Chen, D., Wang, Y., Peng, X. and Zhou, L. (2017). Dynamic buckling of axially functionally-graded beams with non-uniform cross-section under elastic compression stress wave. *Zhendongyu Chongji/Journal of Vibration and Shock 36(13)*, 27-32 and 73.

Elishakoff, I. and Guede, Z. (2004). Analytical polynomial solutions for vibrating axially graded beams. *Mechanics of Advanced Materials and Structures 11*, 517–533.

Elishakoff, I., Kaplunov, J. and Nolde, E. (2015). Celebrating the centenary of Timoshenko's study of effects of shear deformation and rotary inertia. *Applied Mechanics Reviews, 67(6)*, 060802.

Gan, B. S., Huong, T. T. and Kien, N. D., (2017). Post-buckling behavior of axially FGM planar beams and frames. *Procedia Engineering, 171*, 147–158.

Gan, B. S., Trinh, T. H., Le, T. H. and Nguyen, D. K. (2015). Dynamic response of non-uniform Timoshenko beams made of axially FGM subjected to multiple moving point loads. *Structural Engineering and Mechanics, 53(5),* 981-995.

Ghazaryan, D., Burlayenko, V. N., Avetisyan, A. and Bhaskar, A. (2017). Free vibration analysis of functionally graded beams with non-uniform cross-section using the differential transform method. *Journal of Engineering Mathematics,* 1-25.

He, P., Liu, Z and Li, C. (2013), "An improved beam element for beams with variable axial parameters," *Shock and Vibration, 20*, 601-617.

Hein H. and Feklistova L. (2011). Free vibrations of non-uniform and axially functionally graded beams using Haar wavelets. *Engineering Structures, 33(12)*, 3696-3701,

Huang, Y. and Li, X. F. (2010). A new approach for free vibration of axially functionally graded beams with non-uniform cross-section. *Journal of Sound and Vibration 329*, 2291–2303.

Huang, Y., Yang, L. and Luo, Q. (2013). Free vibration of axially functionally graded Timoshenko beams with non-uniform cross-section. *Composites: Part B, 45*, 1493–1498.

Huang, Y., Zhang, M. and Rong H. (2016), Buckling Analysis of Axially Functionally Graded and Non-Uniform Beams Based on Timoshenko Theory. *Acta Mechanica Solida Sinica, 29(2)*, 200-207.

Ilanko S. and Monterrubio L. E. (2014). *The Rayleigh-Ritz Method for Structural Analysis*. London, U.K. John Wiley & Sons Inc.

Karami G., Malekzadeh P. and Shahpari S.A. (2003). DQEM for vibration of shear deformable nonuniform beams with general boundary conditions. *Engineering Structures; 25*, 1169-1178.

Karnovsky I. A. and Lebed O. I. (2004). *Non-classical vibrations of arches and beams*. New York, USA, McGraw-Hill Inc.

Lu, C. F. and Chen, W. Q. (2005). Free vibration of orthotropic functionally graded beams with various end conditions. *Structural Engineering and Mechanics 20*, 465–476.

Mahamood R. M., Akinlabi E. T, Shukla M. and Pityana S. (2012). Functionally Graded Material: An Overview. *Proceedings of the World Congress on Engineering 2012 Vol III WCE 2012, July 4 - 6, 2012, London, U.K.*

Nakamura, T., Wang, T. and Sampath, S. (2000) Determination of properties of graded materials by inverse analysis and instrumented indentation. *Acta Materialia, 48*, 4293–4306.

Niino, M., Hirai, T. and. Watanabe R. (1987). The functionally gradient materials. *J Jap Soc Compos Mat. 13*, 257-264.

Nikolić, A. (2017). Free vibration analysis of a non-uniform axially functionally graded cantilever beam with a tip body. *Archive of Applied Mechanics, 87(7)*, 1227-1241.

Rajasekaran, S. and NorouzzadehTochaei, E. (2014. Free vibration analysis of axially functionally graded tapered Timoshenko beams using differential transformation element method and differential quadrature element method of lowest-order. *Meccanica, 49 (4),* 995-1009.

Reddy J. N. (2005). *An introduction to the finite element method,* New York, USA. McGraw-Hill.

Rossit, C. A., Bambill, D. V. and Gilardi, G. J. (2017). Free vibrations of AFG cantilever tapered beams carrying attached masses. *Structural Engineering and Mechanics, 61(5)*, 685-691.

Rossit, C. A., Bambill, D. V. and Gilardi G. J., (2018). Timoshenko theory effect on the vibration of axially functionally graded cantilever beams carrying concentrated masses. *Structural Engineering and Mechanics*. Accepted for publication.

Sankar, B. V. (2001). An elasticity solution for functionally graded beams. *Composites Science and Technology 61*, 689–696.

Sarkar, K. and Ganguli, R. (2014). Closed-form solutions for axially functionally graded Timoshenko beams having uniform cross-section and fixed–fixed boundary condition. *Composites: Part B, 58*, 361–370.

Shahba, A., Attarnejad, R., Marvi, M. T. and Hajilar, S. (2011). Free vibration and stability analysis of axially functionally graded tapered Timoshenko beams with classical and non-classical boundary conditions. *Composites: Part B, 42*, 801–808.

Shahba, A. and Rajasekaran, S. (2012). Free vibration and stability of tapered Euler-Bernoulli beams made of axially functionally graded materials. *Applied Mathematical modelling 36,* 3094–3111.

Simsek, M., Kocatürk, T. and Akbas, S. D. (2011). Dynamics of an axially functionally graded beam carrying a moving harmonic load. *16th International Conference on Composite Structures Porto-Portugal, June*.

Su H., Banerjee J. R. and Cheung C. W. (2013). Dynamic stiffness formulation and free vibration analysis of functionally graded beams. *Composite Structures, 106*, 854–862.

Tang, A. Y., Wu, J. X., Li, X. F. and Lee, K. Y. (2014). Exact frequency equations of free vibration of exponentially non-uniform functionally graded Timoshenko beams. *International Journal of Mechanical Sciences, 89,* 1-11.

Timoshenko, S. P. (1921). On the correction for shear of the differential equation for transverse vibration of prismatic bar. *Philosophical Magazine 41*, 744-746.

Timoshenko, S. P. (1922). On the transverse vibrations of bars of uniform cross section. *Philosophical Magazine, 43*, 125-131.

Tudjono, S., Han, A., Nguyen, D., Kiryu, S., and Gan, B. (2017). Exact Shape Functions for Timoshenko Beam Element. *IOSR Journal of Computer Engineering (IOSR-JCE), 19(3)*, 12-20.

Wang S. S. (1983). Fracture mechanics for delamination problems in composite materials. *Journal of Composite Materials. 17(3),* 210-223.

Wu L., Wang Q. and Elishakoff I. (2005). Semi-inverse method for axially functionally graded beams with an anti-symmetric vibration mode. *Journal of Sound and Vibration 284*, 1190–1202.

Xie, X., Zheng, H., Zou, X. (2017). An integrated spectral collocation approach for the static and free vibration analyses of axially functionally graded non-uniform beams. *Proceedings of the Institution of Mechanical Engineers, Part C: Journal of Mechanical Engineering Science. 231(13)*, 2459-2471.

Zhao, Y., Huang, Y. and Guo, M., (2017), A novel approach for free vibration of axially functionally graded beams with non-uniform cross-section based on Chebyshev polynomials theory. *Composite Structures, 168*, 277–284.

Zhong, Z. and Yu, T. (2007). Analytical solution of a cantilever functionally graded beam. *Composites Science and Technology 67*, 481–488.

Zong Z. and Zhang Y. Y. (2009). *Advanced differential quadrature methods*. New York, USA. CRC Press.

In: Computational Mechanics (CM)
Editor: Jacob Yuen
ISBN: 978-1-53613-672-2

Chapter 2

A Free Vibration Analysis of Microbeams by Modified Strain Gradient Theory

Diana V. Bambill[1,2,*], Graciela I. Guerrero[2] and Daniel H. Felix[2]

[1]Consejo Nacional de Investigaciones Científicas y Técnicas (CONICET), Bahía Blanca, Argentina

[2]Department of Engineering, Universidad Nacional del Sur (UNS), Bahía Blanca, Argentina

Abstract

The paper presents a model based on the modified strain gradient theory for the free vibration analysis of isotropic microbeams. The analysis includes the effects of rotatory inertia and shear deformation on natural frequencies. It takes into account the dependence of the mechanical properties of the material on the size of the microbeams.

The model is sensitive to the Poisson's ratio and its effect can be observed on the natural frequency coefficients. The numerical results

[*] Corresponding Author Email: dbambill@criba.edu.ar.

developed in this paper include Timoshenko and Bernoulli – Euler microbeams considering different boundary conditions: simply supported–simply supported, clamped–free, clamped–simply supported, clamped–clamped, simply supported– free, free–free. The effects that boundary supports of the microbeams have on the natural non-dimensional frequencies are also studied. Numerical solutions are obtained using the Ritz method with sets of admissible functions that best model the boundary conditions of the microbeam. Additionally, results obtained with the classical theory and with the modified couple stress theory are also presented.

Keywords: microbeam, vibration, Timoshenko, material length scale parameters, modified strain gradient elasticity theory, Bernoulli-Euler, Poisson, Ritz.

INTRODUCTION

Nowadays, the use of Micro ElectroMechanical Systems (MEMS) has been expanded to countless cutting edge technological applications (AZoNano, 2006). MEMS use structural components of micrometric dimensions, among which microbeams are very common. These particular elements are used in important electronic applications, e.g., to create faster processors or memories with greater capacity to store information in increasingly smaller spaces. Moreover, MEMS are used to control the micromirrors to reflect image beams in advanced displays. They also appear as components in the sensors of the accelerometers contained in the automotive airbags or in inkjet printers. Medicine is another field where MEMS are effectively applied. For example, they are present in the special devices that are used to explore the inside of the human body and in disposable blood pressure meters. In several applications, these microstructures act in vibratory environments. Therefore, it is important to know the natural frequencies of the microcomponents.

The size dependence that occurs in the behavior of deformation in microscale structures has been demonstrated experimentally (Yang et al., 2002). In microbeams, where an important dimensional decrease occurs, the size effect has a major role on static and dynamic deformation behavior. It has been shown that resistant structures are stiffened when their size is reduced to micrometric scale.

The classical elasticity theory (CT) describes the behavior of a material at a macrometric scale, while considering as material constants only the Lame's constants (the bulk and shear modules). Furthermore, classical theory underestimates stresses and size dependent behaviors in micro scale structures. Because of this, higher-order continuum theories, that contain additional material length scale parameters, have been used to evaluate the size dependence of microbeams. These theories include the micropolar elasticity, the non-local theory of Eringen, the couple stress, and the strain gradient elasticity theories. The couple stress theory originated from the theory developed by the Cosserat brothers (Cosserat and Cosserat, 1907) and the models proposed by Toupin (1962) and Mindlin (1964). The most novel concept of this couple stress theory is that it introduces a couple stress caused by an independent micro-rotation that is incorporated into the field of displacements. The modified couple stress theory (MCST) proposed by Yang and colleagues (Yang et al., 2002) presents only one additional material constant. The strain gradient elasticity theory adds five additional material constants, considering the contribution of strain gradients. The modified strain gradient theory (MSGT) was proposed by Lam and co- authors (Lam et al., 2003). It enables the application of the higher-order equilibrium conditions to strain gradient elasticity theory and it reduces the number of independent elastic length scale parameters from five to three. The modified couple stress theory considers the structure less rigid than the strain gradient theory (Yang et al., 2002; Ma et al., 2008). These new material constants, as the classical ones, must be determined by means of measurements in experimental tests.

The Timoshenko microbeam based on the couple stress theory has been studied by numerous researchers such as Araújo Dos Santos and Reddy (2012). In 2010 Wang and colleagues (Wang et al., 2010) developed a micro scale Timoshenko beam model based on strain gradient elasticity theory and recovered the micro scale Bernoulli–Euler beam model when shear deformation is ignored. Ansari et al. (2011) investigated free vibration characteristics of functionally graded Timoshenko microbeams based on the strain gradient theory. In 2014, Zhang and others presented a Timoshenko beam element for the analysis of the static bending, free vibration and buckling behaviors of Timoshenko microbeams. The element is based on the strain gradient elasticity theory (Zhang et al., 2014). A bending analysis of micro-sized beams based on the Bernoulli-Euler beam theory is presented by Akgöz et al. (2012), who applied the modified strain gradient elasticity and modified couple stress theories. Besides, they investigated the Poisson's ratio effect. Also Ma et al. (2008), Kahrobaiyan et al. (2012) and Bambill et al. (2016) analyzed Bernoulli-Euler microbeams with non-rigid supports. Kong et al. (2009) analytically solved dynamic problems of Bernoulli–Euler beams based on the strain gradient elasticity theory. Park and Gao (2006) employed a variational formulation based on the principle of minimum total potential energy for the bending of Bernoulli–Euler microbeams.

Akgöz et al. (2013) presented a new size-dependent higher-order-shear-deformation beam model to capture both, the microstructural and the shear deformation effects without using any shear correction factors. The nonlinear forced vibrations of a microbeam, employing the strain gradient elasticity theory, were investigated through the pseudo-arc length continuation technique, obtaining the frequency–response curves of the system (Ghayesh et al., 2013).

Formulation of the Strain Gradient Elasticity Theory

Based on the modified strain gradient theory proposed by Lam et al. (2003), the strain energy of a microbeam depends on both the conventional strain (the symmetric part of the first-order deformation gradient) and on the second-order deformation gradient. Then, the strain energy for a body occupying a volume V can be written as (Zhang et al., 2014)

$$U = \frac{1}{2}\int_V \left(\sigma_{ij}\varepsilon_{ij} + m_{ij}^s \chi_{ij}^s + p_i\gamma_i + \tau_{ijk}^{(i)}\eta_{ijk}^{(i)}\right)dV \tag{1}$$

where σ_{ij} are the classical stresses; m_{ij}^s, p_i and $\tau_{ijk}^{(1)}$ are the higher-order stresses; ε_{ij} is the classical strain tensor; and χ_{ij}^s, γ_i and $\eta_{iijk}^{(1)}$ are the second order strain metrics. V is the volume of the microbeam.

The Cauchy stress tensor (classical stresses) is

$$\sigma_{ij} = \frac{E}{1+\nu}\left[\varepsilon_{ij} + \frac{\nu}{1-2\nu}\varepsilon_{kk}\delta_{ij}\right] - \lambda\varepsilon_{mm}\delta_{ij} + 2\mu\varepsilon_{ij} \text{ with} \tag{2}$$

$$\varepsilon_{ij} = \frac{1}{2}\left(u_{i,j} + u_{j,i}\right) \tag{3}$$

where u_i is the displacement vector; ε_{mm} is the dilation strain and δ_{ij} is the Kronecker symbol (gradient operator). E is the elastic modulus and ν is the Poisson's ratio for isotropic materials.

The bulk and shear modules are λ and μ, they are the conventional Lame's constants (Timoshenko and Goodier, 1970) and appear in the constitutive equation Eq. (2) of the classical stress.

$$\lambda = \frac{E\nu}{(1+\nu)(1-2\nu)}, \quad \mu = G = \frac{E}{2(1+\nu)} \tag{4}$$

The shear modulus or modulus of rigidity is denoted by μ, or sometimes G.

The higher order stresses are defined as

$$m_{ij}^s = 2l_2^{\,2}\mu\chi_{ij}^s \tag{5}$$

$$p_i = 2\mu l_0^2\gamma_i = 2\mu l_0^2\varepsilon_{mm,i} \tag{6}$$

$$\tau_{ijk}^{(1)} = 2\mu l_1^2\eta_{ijk}^{(1)} \tag{7}$$

where l_2, l_0 and l_1 are additional and independent material length scale parameters. In particular, l_2 is associated with the symmetric rotation gradients χ_{ij}^s, l_0 with the dilation gradients γ_i and l_1 with the deviatoric stretch gradients $\eta_{iijk}^{(1)}$.

$$\chi_{ij}^s = \frac{1}{2}\left(\chi_{ij} + \chi_{ji}\right) = \frac{1}{2}(\theta_{i,j} + \theta_{j,i}) \text{, with } \theta_i = \frac{1}{2} e_{ijk} u_{k,j} \tag{8}$$

$$\gamma_i = \varepsilon_{\mathrm{mm},i} \tag{9}$$

$$\begin{aligned} \eta_{ijk}^{(1)} &= \frac{1}{3}\left(\varepsilon_{jk,i} + \varepsilon_{ki,j} + \varepsilon_{ij,k}\right) - \frac{1}{15} \\ &\left[\delta_{ij}\left(\varepsilon_{mm,k} + 2\varepsilon_{mk,\mathrm{m}}\right) + \delta_{jk}\left(\varepsilon_{mm,\mathrm{i}} + 2\varepsilon_{mi,\mathrm{m}}\right) + \delta_{ki}\left(\varepsilon_{mm,\mathrm{j}} + 2\varepsilon_{mj,\mathrm{m}}\right)\right] \end{aligned} \tag{10}$$

where the permutation symbol is e_{ijk}.

ENERGY OF THE VIBRATING TIMOSHENKO MICROBEAM

The microbeam model is illustrated in Figure 1. L is the length of the microbeam of rectangular cross section. The origin of the Cartesian coordinate system is taken at the left end of the beam and the $\bar{x}$-axis coincides with the centroidal axis of the microbeam. The vibrating beam translates in the $\bar{z}$ - direction. Six different classical boundary conditions are schematized in the figure.

Timoshenko beam theory takes into account the rotational inertia, the shear deformation, and its combined effects. The displacements are the transverse displacement $w(\bar{x},t)$ in the $\bar{z}$-direction and the angle $\phi(\bar{x},t)$, which is the rotation angle of a transverse normal about the $\bar{y}$-axis, (Figure 2). t is the time. The considered displacements are due to bending and shear deformation.

For the analysis, microbeams with constant cross section A are considered. Then:

$$dV = dA\, d\bar{x} \tag{11}$$

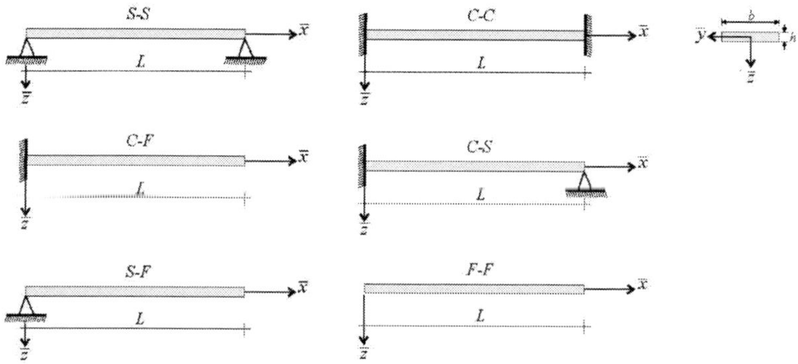

Figure 1. Microbeams configurations.

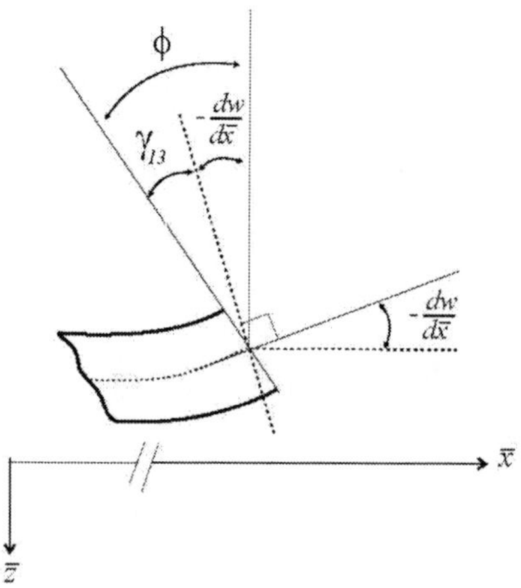

Figure 2. Microbeam kinematic assumptions.

The strain energy U of the microbeam from Eq. (1) is defined by

$$U = \frac{1}{2}\int_L d\,\bar{x}\Big[\int_A \left(\sigma_{11}\varepsilon_{11} + 2\sigma_{13}\varepsilon_{13} + 2m_{12}^{s}\chi_{12}^{s} + p_1\gamma_1 + p_3\gamma_3 + \right. \\ \left. + \tau_{111}^{(1)}\eta_{111}^{(1)} + 3\tau_{131}^{(1)}\eta_{131}^{(1)} + 6\tau_{221}^{(1)}\eta_{221}^{(1)} + 3\tau_{232}^{(1)}\eta_{232}^{(1)} + \tau_{333}^{(1)}\eta_{333}^{(1)}\right) dA\Big] \tag{12}$$

The corresponding displacement field is

$$
\begin{aligned}
u(\bar{x},t) &= u_1(\bar{x},t) = 0 - z\phi(\bar{x},t) \\
v(\bar{x},t) &= u_2(\bar{x},t) = 0 \\
w(\bar{x},t) &= u_3(\bar{x},t)
\end{aligned}
\tag{13}
$$

Note that in the Timoshenko beam theory $\phi(\bar{x},t) \neq \frac{\partial w(\bar{x},t)}{\partial x}$, the planar cross sections are initially perpendicular to the neutral axis of the microbeam, but due to shear deformation they do not remain perpendicular to the neutral axis during deflection (Timoshenko, 1921; Araújo Dos Santos and Reddy, 2012).

In terms of the displacements, the strain gradient elasticity theory applied to the Timoshenko microbeam can be analyzed in the following way:

Strains and Stresses of the Classical Elastic Theory

Replacing Eqs. (13) into Eq. (3) the components of the classical strain tensor are:

$$
\begin{aligned}
\varepsilon_{11} &= -z\frac{\partial \phi(\bar{x},t)}{\partial \bar{x}} \\
\varepsilon_{13} &= \frac{1}{2}\left(\frac{\partial w(\bar{x},t)}{\partial \bar{x}} - \phi(\bar{x},t)\right)
\end{aligned}
\tag{14}
$$

and the classical stresses Eq. (2) results in:

$$
\begin{aligned}
\sigma_{11} &= -(\lambda + 2\mu)z\frac{\partial \phi(\bar{x},t)}{\partial \bar{x}} \\
\sigma_{22} = \sigma_{33} &= -\lambda z\frac{\partial \phi(\bar{x},t)}{\partial \bar{x}} \\
\sigma_{13} &= \mu\left(\frac{\partial w(\bar{x},t)}{\partial \bar{x}} - \phi(\bar{x},t)\right)
\end{aligned}
\tag{15}
$$

The stress components σ_{22} and σ_{33} appear because of Poisson's effect.

Strains and Stresses of the Rotation Gradient

From Eqs. (8) the strains become:

$$\begin{aligned}\chi_{12}^{s} &= -\frac{1}{4}\left(\frac{\partial^2 w(\overline{x},t)}{\partial\overline{x}^2}+\frac{\partial\phi(\overline{x},t)}{\partial\overline{x}}\right)\\ \chi_{21}^{s} &= -\frac{1}{4}\left(\frac{\partial^2 w(\overline{x},t)}{\partial\overline{x}^2}+\frac{\partial\phi(\overline{x},t)}{\partial\overline{x}}\right)\end{aligned} \tag{16}$$

and the couple stresses, Eq. (5), are defined as

$$\begin{aligned}m_{12}^{s} &= -\frac{1}{2}\mu l_2^2\left(\frac{\partial^2 w(\overline{x},t)}{\partial\overline{x}^2}+\frac{\partial\phi(\overline{x},t)}{\partial\overline{x}}\right)\\ m_{21}^{s} &= -\frac{1}{2}\mu l_2^2\left(\frac{\partial^2 w(\overline{x},t)}{\partial\overline{x}^2}+\frac{\partial\phi(\overline{x},t)}{\partial\overline{x}}\right)\end{aligned} \tag{17}$$

Strains and Stresses of the Dilation Gradient

The components of strain by Eq. (9) assumed the form:

$$\begin{aligned}\gamma_1 &= -z\frac{\partial^2\phi(\overline{x},t)}{\partial\overline{x}^2}\\ \gamma_3 &= -\frac{\partial\phi(\overline{x},t)}{\partial\overline{x}}\end{aligned} \tag{18}$$

and from Eq. (6), the dilation stresses become:

$$\begin{aligned}p_1 &= -2\mu l_0^2 z\frac{\partial^2\phi(\overline{x},t)}{\partial\overline{x}^2}\\ p_3 &= -2\mu l_0^2\frac{\partial\phi(\overline{x},t)}{\partial\overline{x}}\end{aligned} \tag{19}$$

Stains and Stresses of the Higher Order Deviatoric Stretch Gradient

The strain components given by Eq. (10) are:

$$\begin{aligned}
&\eta_{111}^{(1)} = -\frac{2}{5} z \frac{\partial\phi^2(\overline{x},t)}{\partial\overline{x}^2} \\
&\eta_{131}^{(1)} = \eta_{113}^{(1)} = \eta_{311}^{(1)} = -\frac{8}{15}\frac{\partial\phi(\overline{x},t)}{\partial\overline{x}} + \frac{4}{15}\frac{\partial^2 w(\overline{x},t)}{\partial\overline{x}^2} \\
&\eta_{133}^{(1)} = \eta_{212}^{(1)} = \eta_{221}^{(1)} = \eta_{313}^{(1)} = \eta_{331}^{(1)} = \eta_{122}^{(1)} = \frac{1}{5} z \frac{\partial^2\phi(\overline{x},t)}{\partial\overline{x}^2} \\
&\eta_{223}^{(1)} = \eta_{232}^{(1)} = \eta_{322}^{(1)} = -\frac{1}{15}\left(-2\frac{\partial\phi(\overline{x},t)}{\partial\overline{x}} + \frac{\partial^2 w(\overline{x},t)}{\partial\overline{x}^2}\right) \\
&\eta_{333}^{(1)} = -\frac{1}{5}\left(-2\frac{\partial\phi(\overline{x},t)}{\partial\overline{x}} + \frac{\partial^2 w(\overline{x},t)}{\partial\overline{x}^2}\right)
\end{aligned} \tag{20}$$

and the stresses of the deviatoric tensor, Eq. (7), become:

$$\begin{aligned}
&\tau_{111}^{(1)} = -\frac{4}{5}\mu l_1^2 z \frac{\partial\phi^2(\overline{x},t)}{\partial\overline{x}^2} \\
&\tau_{131}^{(1)} = \tau_{113}^{(1)} = \tau_{311}^{(1)} = \frac{8}{15}\mu l_1^2\left(-2\frac{\partial\phi(\overline{x},t)}{\partial\overline{x}} + \frac{\partial^2 w(\overline{x},t)}{\partial\overline{x}^2}\right) \\
&\tau_{133}^{(1)} = \tau_{212}^{(1)} = \tau_{221}^{(1)} = \tau_{313}^{(1)} = \tau_{331}^{(1)} = \tau_{122}^{(1)} = \frac{2}{5}\mu l_1^2 z \frac{\partial^2\phi(\overline{x},t)}{\partial\overline{x}^2} \\
&\tau_{223}^{(1)} = \tau_{232}^{(1)} = \tau_{322}^{(1)} = -\frac{2}{15}\mu l_1^2\left(-2\frac{\partial\phi(\overline{x},t)}{\partial\overline{x}} + \frac{\partial^2 w(\overline{x},t)}{\partial\overline{x}^2}\right) \\
&\tau_{333}^{(1)} = -\frac{2}{5}\mu l_1^2\left(-2\frac{\partial\phi(\overline{x},t)}{\partial\overline{x}} + \frac{\partial^2 w(\overline{x},t)}{\partial\overline{x}^2}\right)
\end{aligned} \tag{21}$$

Replacing Eqs. (14-21) in Eq. (12), the strain energy of a microbeam made of isotropic elastic material results in a function that depends on the symmetric strain tensor, the dilation gradient vector, the deviatoric stretch gradient tensor, and the symmetric rotation gradient (Lam et al., 2003). The total strain energy of a microbeam in terms of the modified strain gradient elasticity theory (MSGT) is

$$U = \frac{1}{2}\int_0^L \{ \ (\lambda + 2\mu) I\phi'^2 + \mu\kappa_s A(w' - \phi)^2 + \frac{\mu}{4} l_2^2 A(w'' + \phi')^2 + \\ +2\mu l_0^2 A\,\phi'^2 + (2\mu l_0^2 I + \frac{4}{5}\mu l_1^2 I)\phi''^2 + \frac{8}{15}\mu l_1^2 A(w'' - 2\phi')^2 \ \} d\bar{x} \tag{22}$$

where $(\ )'$ denotes the derivative with respect to $\bar{x}$; κ_s is the shear correction factor related to the area of the cross section A and $I = I_y = \int_A z^2 dA$ is the second moment of area or moment of inertia. This theory has three additional material constants: l_0, l_1, l_2, which take into account the size dependency.

In particular, the classical elasticity-strain-energy terms due to bending and shear are:

$$U^{CT} = \frac{1}{2}\int_0^L \{ \ (\lambda + 2\mu) I\phi'^2 + \mu\kappa_s A(w' - \phi)^2 \ \} d\bar{x} \tag{23}$$

The energy term that corresponds to the couple stress effect is:

$$U^{MCST} = \frac{1}{2}\int_0^L \{ \ \frac{\mu}{4} l_2^2 A(w'' + \phi')^2 \ \} d\bar{x} \tag{24}$$

And the energy terms that correspond to the effects of the dilation gradient, the rotation gradient and the deviatoric stretch gradient are expressed as:

$$U^{MSGT} = \frac{1}{2}\int_0^L \{ \ 2\mu l_0^2 A\,\phi'^2 + (2\mu l_0^2 I + \frac{4}{5}\mu l_1^2 I)\phi''^2 + \frac{8}{15}\mu l_1^2 A(w'' - 2\phi')^2 \ \} d\bar{x} \tag{25}$$

Finally considering the displacement field components, Eqs. (13), the kinetic energy of the homogeneous microbeam in terms of displacements is

$$T = \frac{1}{2}\int_0^L \left[\rho A\left(\frac{\partial w(\bar{x},t)}{\partial t}\right)^2 + \rho I\left(\frac{\partial \phi(\bar{x},t)}{\partial t}\right)^2\right] d\bar{x} = \\ = \frac{1}{2}\int_0^L \left[\rho A(\dot{w})^2 + \rho I(\dot{\phi})^2\right] d\bar{x} \tag{26}$$

where the translational inertia is ρA and the rotatory inertia is ρI, with ρ the material density.

For the normal modes of vibration of the microbeam, the following solutions are proposed

$$w(\bar{x},t) = W(\bar{x})\, e^{i\omega t} \tag{27}$$

$$\phi(\bar{x},t) = \varphi(\bar{x}) e^{i\omega t} \tag{28}$$

where $i = \sqrt{-1}$ and the natural circular frequency expressed in radians per second is ω.

Substituting Eqs. (27-28) in Eqs. (23-26) the expressions for the potential and kinetic energy are given by:

$$U = \frac{e^{2i\omega t}}{2}\left\{\int_0^L \left\{ (\lambda + 2\mu) I\varphi'^2 + \mu\kappa_s A(W' - \varphi)^2 + \frac{\mu}{4} l_2^2 A(W'' + \varphi')^2 + \right.\right. \\ \left.\left. + 2\mu l_0^2 A\varphi'^2 + (2\mu l_0^2 I + \frac{4}{5}\mu l_1^2 I)\varphi''^2 + \frac{8}{15}\mu l_1^2 A(W'' - 2\varphi')^2 \right\} d\bar{x} \right\} \tag{29}$$

$$T = e^{2i\omega t}\omega^2 \frac{\rho}{2}\int_0^L \left[A(W)^2 + I(\varphi)^2\right] d\bar{x} \tag{30}$$

RITZ METHOD

The Ritz method is a numerical method that allows to obtain an approximate solution of the natural frequencies. For the sake of clarity, the non-dimensional coordinate x is adopted:

$$x = \frac{\bar{x}}{L}; \text{ with } x \in [0,1] \text{ while } \bar{x} \in [0, L] \quad (31)$$

According to this method, it is necessary to approximate the maximum transversal displacement and rotation angle of the deflected shape of the microbeam. Then, the following summations of products of undetermined coefficients and admissible functions approximate the maximum transversal displacement and rotation angle (Ilanko and Monterrubio 2014).

$$W(x) \cong W_a(x) = \sum_{i=1}^{n} C_i.W_i(x) \quad (32)$$

$$\varphi(x) \cong \varphi_a(x) = \sum_{j=1}^{n} D_j.\varphi_j(x) \quad (33)$$

where n is the number of terms; $W_i(x)$ and $\varphi_i(x)$ are the admissible functions; C_i and D_i, are the undetermined coefficients.

The admissible functions satisfy the essential boundary conditions of the microbeam and represent an approximate shape of the deflected vibrating microbeam. They are

$$W_i(x) = x^{i-1}.f_W(x) \text{ with } i = 1,2,...n. \quad (34)$$

$$\varphi_j(x) = x^{j-1}.f_\varphi(x) \text{ with } j = 1,2,...n. \quad (35)$$

where $f_W(x)$ and $f_\varphi(x)$ are polynomial functions of the form

$$f_W(x) = a_0 + a_1 . x + a_2 . x^2 \tag{36}$$

$$f_\varphi(x) = b_0 + b_1 . x + b_2 . x^2 \tag{37}$$

and $a_0, a_1, a_2, b_0, b_1, b_2$ are constants, which are determined by applying the boundary conditions of the beam and depend on the nature of each edge support. The boundary conditions of the microbeam are assumed neglecting higher order effects (Zhang et al., 2014). For example, for classical boundary conditions, the following admissible functions (Figures 3-8) are proposed

Simply Supported –Simply Supported Microbeam (S-S)

$$a_0 = 0;\ a_1 = -1;\ a_2 = 1;\ b_0 = 1;\ b_1 = 0;\ b_2 = 0 \tag{38}$$

$$f_W(x) = -x + x^2 \tag{39}$$

$$f_\varphi(x) = 1 \tag{40}$$

Clamped-Clamped Microbeam (C-C)

$$a_0 = 0;\ a_1 = -1;\ a_2 = 1;\ b_0 = 0;\ b_1 = 1;\ b_2 = -1 \tag{41}$$

$$f_W(x) = -x + x^2 \tag{42}$$

$$f_\varphi(x) = x - x^2 \tag{43}$$

Clamped-Free Microbeam (C-F)

$$a_0 = 0;\ a_1 = 1;\ a_2 = 1;\ b_0 = 0;\ b_1 = 1;\ b_2 = -\frac{1}{2} \tag{44}$$

$$f_W(x) = x + x^2 \tag{45}$$

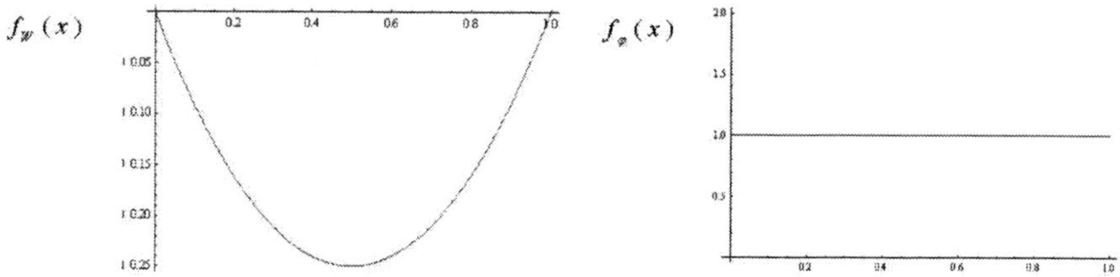

Figure 3. Admissible functions for S-S microbeams.

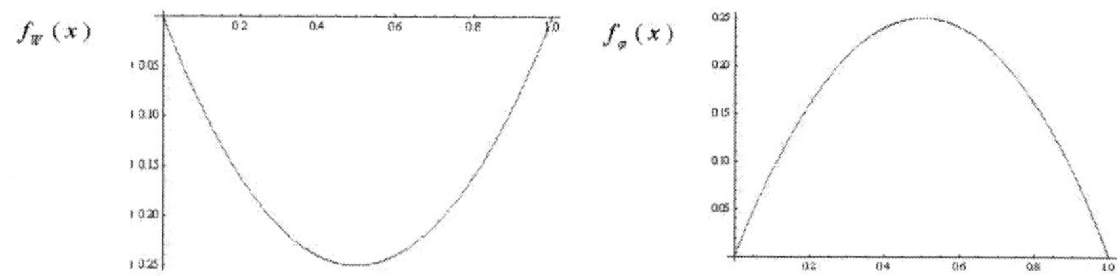

Figure 4. Admissible functions for C-C microbeams.

$$f_\varphi(x) = x - \frac{1}{2}x^2 \tag{46}$$

Clamped-Simply Supported Microbeam (C-S)

$$a_0 = 0;\ a_1 = -1;\ a_2 = 1;\ b_0 = 0;\ b_1 = 1;\ b_2 = -\frac{1}{2} \tag{47}$$

$$f_W(x) = -x + x^2 \tag{48}$$

$$f_\varphi(x) = x - \frac{1}{2}x^2 \tag{49}$$

Simply Supported –Free Microbeam (S-F), (Unstable).

$$a_0 = 0;\ a_1 = 1;\ a_2 = 1;\ b_0 = 1;\ b_1 = 0;\ b_2 = 0 \tag{50}$$

$$f_W(x) = x + x^2 \tag{51}$$

$$f_\varphi(x) = 1 \tag{52}$$

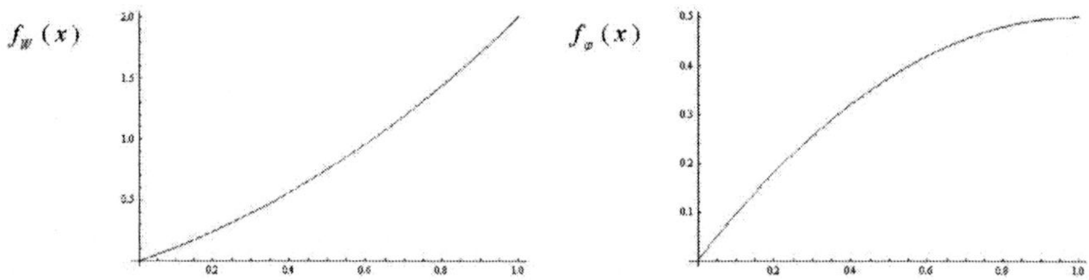

Figure 5. Admissible functions for C-F microbeams.

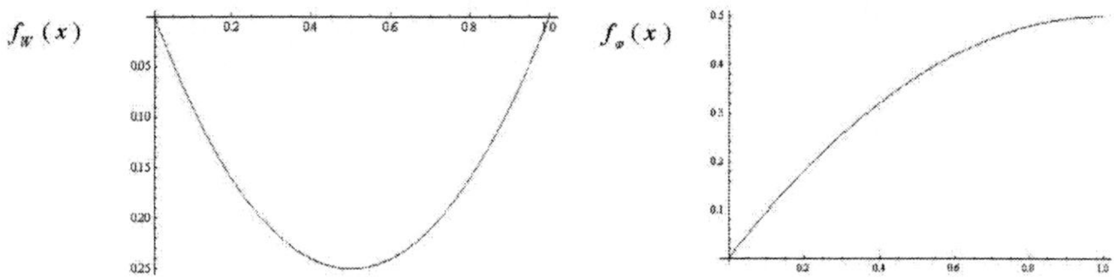

Figure 6. Admissible functions for C-S microbeams.

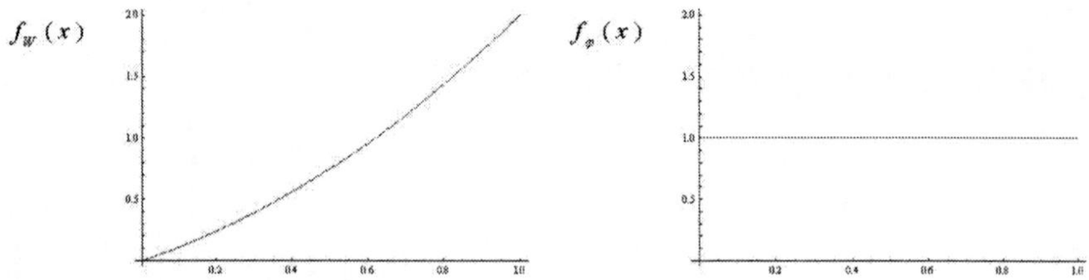

Figure 7. Admissible functions for S-F microbeams.

Free-Free Microbeam (F-F), (Unstable)

$$a_0 = 1;\ a_1 = 0;\ a_2 = 0;\ b_0 = 1;\ b_1 = 0;\ b_2 = 0 \tag{53}$$

$$f_W(x) = 1 \tag{54}$$

$$f_\varphi(x) = 1 \tag{55}$$

The energy functional J for the vibrating homogeneous microbeam is

$$\begin{aligned} J &= U_{\max} - T_{\max} \\ &= \frac{1}{2}\left\{\int_0^1 \left\{\left[(\lambda+2\mu) I \frac{\varphi_a'^2}{L^2}\right] + \mu\kappa_s A\left(W_a' - \varphi_a\right)^2 + \frac{\mu}{4} l_2^2 A \frac{\left(W_a'' + \varphi_a'\right)^2}{L^2} + \right.\right. \\ &\left.\left. + 2\mu l_0^2 A \frac{\varphi_a'^2}{L^2} + \left(2\mu l_0^2 I + \frac{4}{5}\mu l_1^2 I\right)\frac{\varphi_a''^2}{L^4} + \frac{8}{15}\mu l_1^2 A \frac{\left(W_a'' - 2\varphi_a'\right)^2}{L^2}\right\} L dx\right\} \\ &- \omega^2 \frac{\rho}{2}\int_0^1 \left[A\left(W_a\right)^2 L^2 + I\left(\varphi_a\right)^2\right] L dx \end{aligned} \tag{56}$$

where $U_{\max}$ is the maximum strain energy and $T_{\max}$. is the maximum kinetic energy.

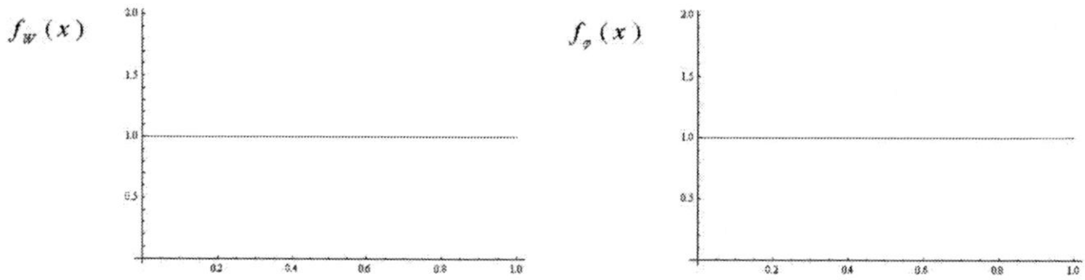

Figure 8. Admissible functions for F-S microbeams.

Following Ritz method, the functional J is minimized with respect to every arbitrary constant as follows

$$\frac{\partial J}{\partial C_i}=0;\quad i=1,2,\ldots,n.$$
$$\frac{\partial J}{\partial D_i}=0;\quad i=1,2,\ldots,n. \tag{57}$$

Then, a linear system of equations is obtained and its non-trivial solution results in the following eigenvalue equation:

$$[\mathbf{K}]\{Q\}-[\mathbf{M}]\{Q\}\Omega^2=\mathbf{0} \tag{58}$$

where $\Omega=\sqrt{\dfrac{\rho A\omega^2 L^4}{EI}}$ corresponds to the non-dimensional frequency coefficients. $\{Q\}$ contains the arbitrary constants C_i and D_i. $[\mathbf{K}]$ is the stiffness matrix and $[\mathbf{M}]$ is mass matrix associated with the inertia.

The stiffness matrix is

$$[\mathbf{K}]=\begin{bmatrix} k_{ii} & k_{ij} \\ k_{ji} & k_{jj} \end{bmatrix} \tag{59}$$

where the interior elements are

$$[k_{11}]_{ii}=\int_0^1\frac{1}{EI}\left[\mu\kappa_s AL^2 W_i' W_j' + \left(\mu A\frac{l_2^2}{4}+\frac{8}{15}\mu A l_1^2\right)W_i'' W_j''\right]dx$$
$$[k_{12}]_{ij}=\int_0^1\frac{1}{EI}\left[-\mu\kappa_s AL^2 \varphi_i W_j' + \left(\mu A\frac{l_2^2}{4}-\frac{16}{15}\mu A l_1^2\right)\varphi_i' W_j''\right]dx$$
$$[k_{21}]_{ij}=\int_0^1\frac{1}{EI}\left[-\mu\kappa_s AL^2 W_i' \varphi_j + \left(\mu A\frac{l_2^2}{4}-\frac{16}{15}\mu A l_1^2\right)W_i'' \varphi_j'\right]dx \tag{60}$$
$$[k_{22}]_{ij}=\int_0^1\frac{1}{EI}\left[\left((\lambda+2\mu)I+\mu A\left(2l_0^2+\frac{l_2^2}{4}+\frac{32}{15}l_1^2\right)\right)\varphi_i' \varphi_j' + \right.$$
$$\left. +\mu\kappa_s AL^2\varphi_i\,\varphi_j + \left(2\mu I l_0^2+\frac{4}{5}\mu I l_1^2\right)\varphi_i'' \varphi_j''\right]dx$$

with $[k_{12}] = [k_{21}]^T$.

The mass matrix

$$[\mathbf{M}] = \begin{bmatrix} g_{ii} & g_{ij} \\ g_{ji} & g_{jj} \end{bmatrix} \tag{61}$$

is built with the following terms.

$$\begin{aligned} (g_{11})_{ij} &= \int_0^1 W_i\, W_j dx \\ (g_{12})_{ij} &= (g_{21})_{ij} = 0 \\ (g_{22})_{ij} &= \int_0^1 \frac{I}{A} \varphi_i\, \varphi_j dx \end{aligned} \tag{62}$$

As it is known, the eigenvalues obtained with the Ritz method are upper bounds of the sought values.

NUMERICAL RESULTS

Free vibrations of microbeams with different combinations of classical boundary conditions are investigated. The microbeam has rectangular cross section. Figure 1. Numerical results using the Ritz method were obtained for the first five eigenvalues.

$$A = b\,h,\ I = \frac{b\,h^3}{12} \tag{63}$$

where h is the thickness and b is the width.

Two new parameters are defined as:

- L/h is the length-to-thickness ratio and is related to slenderness of the microbeam;

- h/l is the thickness-to-material-length scale parameter ratio and is related to the microscale dimensions.

Table 1. MSGT convergence analysis of frequency coefficients of S-S microbeams. $\kappa_s = \frac{5}{6}; \nu = 0.38; \frac{L}{h} = 10$

h/l	n	Ω_1	Ω_2	Ω_3	Ω_4	Ω_5	
1000	5	13.1269	48.8499	126.622	217.210	340.541	Present
	7	13.1232	48.6760	100.082	161.349	279.405	
	9	13.1232	48.6753	98.9040	157.262	228.058	
	11	13.1232	48.6753	98.8869	157.132	219.798	
	12	13.1232	48.6753	98.8869	157.132	219.320	
	13	13.1232	48.6753	98.8869	157.131	219.312	
		13.1232	48.6751	98.8865	-	-	Akgöz et al. (2013)
		13.1232	48.6751	98.9965	-	-	Felix et al. (2017)
100	5	13.1320	48.8672	126.652	217.268	340.654	Present
	7	13.1283	48.6931	100.115	161.400	279.458	
	9	13.1283	48.6924	98.9350	157.308	228.130	
	11	13.1283	48.6924	98.9179	157.178	219.860	
	12	13.1283	48.6924	98.9179	157.178	219.380	
	13	13.1283	48.6924	98.9179	157.176	219.375	
		13.1283	48.6924	98.9179	-	-	Felix et al. (2017)
2	5	21.9436	75.4168	168.083	322.239	555.670	Present
	7	21.9307	75.0327	146.076	232.930	374.549	
	9	21.9299	75.0305	144.085	224.876	330.358	
	11	21.9294	75.0277	144.082	224.712	317.961	
	12	21.9293	75.0277	144.076	224.711	317.389	
	13	21.9293	75.0272	144.076	224.699	317.379	
		21.9293	75.0271	144.0755	-	-	Felix et al. (2017)
1	5	34.1973	106.598	223.991	513.371	659.380	Present
	6	34.1778	106.513	205.784	368.531	659.380	
	7	34.1778	106.118	204.546	340.170	559.654	
	9	34.1763	106.115	202.493	327.141	498.649	
	10	34.1756	106.115	202.491	326.936	484.182	
	11	34.1756	106.112	202.491	326.825	482.999	
	12	34.1755	106.112	202.486	326.823	482.264	

h/l	n	Ω_1	Ω_2	Ω_3	Ω_4	Ω_5	
	13	34.1755	106.111	202.486	326.811	482.238	Present
		34.1755	106.1113	-	-	-	Zhang et al. (2014)
		34.1755	106.1114	202.4855	-	-	Felix et al. (2017)
1/2	5	52.0444	161.572	356.132	659.380	947.230	Present
	7	52.0242	160.861	331.847	588.487	659.380	
	9	52.0226	160.857	329.296	563.369	659.380	
	11	52.0219	160.854	329.293	562.702	659.380	
	12	52.0218	160.854	329.287	562.696	659.380	
	13	52.0218	160.853	329.287	562.680	659.380	

Table 1 presents the convergence results with respect to the number of terms n considered in Eqs. (32-33). The Ritz procedure is performed under some simplifying assumptions on a microbeam simply supported at both ends (S-S), using the MSGT. The parameters are adopted as follows: $L/h = 10$, all the material length scale parameters are considered equal to l ($l_0 = l_1 = l_2 = l$), while the ratio h/l assumes several values. The shear correction factor κ_s is equal to $5/6$ and the adopted Poisson's ratio is $\nu = 0.38$. Table 1 demonstrates monotonic convergence results as is expected when using the proposed method. Particular results are compared with Akgöz et al. (2013), Zhang et al. (2014) and Felix et al. (2017).

All the following results were obtained using $n = 11$ in Eqs. (32) and (33). For solutions of the modified strain gradient theory, all the material length scale parameters are considered equal to l, then $l_0 = l_1 = l_2 = l$. The modified couple stress theory and the classical elasticity theory can be considered as special cases of the higher-order stress theory. In the modified couple stress theory, MCST, the effects of the dilation gradient, (l_0=0) and the deviatoric stretch gradient (l_1=0) are assumed to be negligible (Lam et al., 2003). Besides, if all the material length scale parameters are assumed to be equal to zero, the strain gradient model turns into the classical elasticity model. The CT does not consider the effects of the dilation gradient ($l_0 = 0$), the deviatoric stretch gradient ($l_1 = 0$) and the rotation gradient ($l_2 = 0$).

Table 2. Fundamental frequency coefficients of S-S microbeams.

$$\kappa_s = \frac{5}{6}; \nu = 0.38; \frac{L}{h} = 10$$

h/l	MSGT	MCST	CT	
1000	13.1232	13.1232	13.1232	Present
100	13.1283	13.1247	13.1232	
	13.1283	13.1247	13.1232	Felix et al. (2017)
10	13.6273	13.4801	13.1232	Present
	13.6274	13.2748	13.1232	Akgöz et al. (2013)
5	15.0176	13.7192	13.1232	Present
	15.0180	13.7192	13.1232	Akgöz et al. (2013)
	15.0180	13.7192	13.1232	Zhang et al. (2014)
	15.0175	13.7192	13.1232	Felix et al. (2017)
2	21.9293	16.4813	13.1232	Present
	21.9293	16.4813	13.1232	Felix et al. (2017)
1	34.1755	23.7053	13.1232	Present
	34.1822	23.7053	13.1232	Akgöz et al. (2013)
	34.1822	23.7053	13.1232	Zhang et al. (2014)
	34.1755	23.7053	13.1232	Felix et al. (2017)
$1/2$	52.0218	40.1927	13.1232	Present
$1/3$	66.6014	55.7438	13.1232	
$1/5$	94.3429	78.0201	13.1232	
$1/8$	134.666	95.4679	13.1232	
$1/12$	202.125	108.613	13.1232	

Tables 2 to 6 present results for simply supported-simply supported Timoshenko microbeams based on the described theories of elasticity. The thickness-to-material length scale parameter ratio varies from $h = 1000\,l$ to $h = 0.08\widehat{3}\,l$ (from macro to microscale). The first five natural frequency coefficients are compared with available results of particular values of h/l.

Table 3. Second frequency coefficients of S-S microbeams.

$$\kappa_s = 5/6; \nu = 0.38; L/h = 10$$

h/l	MSGT	MCST	CT	
1000	48.6752	48.6752	48.6751	Present
100	48.6924	48.6809	48.6751	
	48.6924	48.6809	48.6751	Felix et al. (2017)
10	50.3497	49.2437	48.6751	Present
	50.3506	49.2437	48.6751	Akgöz et al. (2013)
5	54.8513	50.9056	48.6751	Present
	54.8583	50.9056	48.6751	Akgöz et al. (2013)
	54.8583	50.9056	48.6751	Zhang et al. (2014)
	54.8512	50.9056	48.6751	Felix et al. (2017)
2	75.0271	61.0896	48.6751	Present
	75.0271	61.0896	48.6751	Felix et al. (2017)
1	106.111	86.4924	48.6751	Present
	106.1615	86.4924	48.6751	Akgöz et al. (2013)
	106.1615	86.4924	48.6751	Zhang et al. (2014)
	106.1114	86.4924	48.6751	Felix et al. (2017)
1/2	160.854	1136.5900	48.6751	Present
1/3	220.452	172.658	48.6751	
1/5	341.674	207.673	48.6751	
1/8	533.301	225.311	48.6751	
1/12	659.380	234.875	48.6751	

The frequency coefficients based on the MSGT and the MCST increase as the ratio of the thickness-to-material-length scale parameter h/l decreases. As the MCST considers the structure less rigid than the MSGT, the MCST coefficients are lower than the corresponding MSGT coefficients. It can be observed that in microbeams that have a thickness more than ten times larger than the material length scale parameter, the effects of the size dependency are very small. In particular, for $h = 1000l$ the MSGT and MCST coefficients tend to those obtained using the classical elasticity theory (CT).

Finally, it is noted that frequency coefficients that are based on the classical elasticity theory are independent of thickness-to-material length scale parameter ratio.

Table 4. Third frequency coefficients of S-S microbeams.

$$\kappa_s = 5/6;\ \nu = 0.38;\ L/h = 10$$

h/l	**MSGT**	**MCST**	**CT**	
1000	98.8869	98.8867	98.8865	Present
100	98.9179	98.8986	98.8865	
	98.9179	98.8986	98.8865	Felix et al. (2017)
10	101.907	100.082	98.8865	Present
	101.9104	100.0818	98.8865	Akgöz et al. (2013)
5	109.865	103.564	98.8865	Present
	109.8844	103.5636	98.8865	Akgöz et al. (2013)
	109.8587	103.5636	98.8865	Felix et al. (2017)
2	144.076	124.533	98.8865	Present
	144.0755	124.5333	98.8865	Felix et al. (2017)
1	202.485	174.072	98.8865	Present
	202.6362	174.0724	98.8865	Akgöz et al. (2013)
	202.4855	174.0725	98.8865	Felix et al. (2017)
1/2	329.287	257.739	98.8865	Present
1/3	471.118	304.448	98.8865	
1/5	659.380	339.492	98.8865	
1/8	659.380	353.946	98.8865	
1/12	824.695	361.040	98.8865	

For particular cases of Tables 1, 2, 3 and 4, results are compared with the analytical solutions given by Akgöz et al. (2013) and with the finite element analysis made by Zhang et al. (2014) and Felix et al. (2017). Excellent agreement is obtained between the present values and those ones provided in references.

Table 5. Fourth frequency coefficients of S-S microbeams.

$\kappa_s = 5/6; \nu = 0.38; L/h = 10$

h/l	MSGT	MCST	CT
1000	157.132	157.132	157.1320
100	157.176	157.151	157.1320
10	161.553	159.148	157.1320
5	202.471	165.005	157.1320
2	224.698	199.650	157.1320
1	326.811	277.166	157.1320
1/2	562.680	389.874	157.1320
1/3	659.380	440.283	157.1320
1/5	753.139	472.339	157.1320
1/8	1145.72	484.313	157.1320
1/12	1852.12	489.957	157.1320

Table 6. Fifth frequency coefficients of S-S microbeams.

$\kappa_s = 5/6; \nu = 0.38; L/h = 10$

h/l	MSGT	MCST	CT
1000	219.798	219.797	219.7970
100	219.384	219.331	219.797
10	225.357	222.364	219.797
5	241.471	231.213	219.797
2	317.363	282.596	219.797
1	482.206	391.545	219.797
1/2	659.380	528.454	219.797
1/3	822.343	578.923	219.797
1/5	1331.46	607.693	219.797
1/8	2035.03	617.876	219.797
1/12	3296.99	622.602	219.797

Table 7 gives the first five frequency coefficients of SS-SS Timoshenko microbeams, $L = 20h$. These results correspond to the modified strain gradient theory with $l_0 = l_1 = l_2 = l$, $\kappa_s = 5/6$ and $\nu = 0.38$.

As expected, the coefficients of more slender microbeams are lower than those that correspond to stubby microbeams, $L = 10\,h$. (Tables 1 to 6).

Table 7. Frequency coefficients of S-S microbeams. MSGT.

$$\kappa_s = 5/6;\ \nu = 0.38;\ L/h = 20$$

h/l	Ω_1	Ω_2	Ω_3	Ω_4	Ω_5
1000	13.4050	52.4929	114.260	194.704	290.720
100	13.4104	52.5133	114.302	194.773	290.818
10	13.9377	54.5093	118.434	201.407	300.107
5	15.4200	60.0714	129.808	219.433	325.063
2	23.1302	87.7253	183.317	300.206	433.397
1	38.7661	136.717	268.524	424.576	605.578
1/2	66.3325	208.104	398.493	643.557	950.138

Table 8. Fundamental frequency coefficients of C-C microbeams.

$$\kappa_s = 5/6;\ \nu = 0.38;\ L/h = 10$$

h/l	**MSGT**	**MCST**	**CT**	
1000	27.2166	27.2165	27.2165	Present
100	27.2257	27.2192	27.2165	
	27.2257	27.2192	27.2165	Felix et al. (2017)
10	28.1075	27.4737	27.2165	Present
5	30.4712	28.1727	27.2165	
	30.4711	28.1716	27.2165	Felix et al. (2017)
2	40.336	31.751	27.2165	Present
	40.3360	31.7506	27.2165	Zhang et al. (2014)
	40.3360	31.7516	27.2165	Felix et al. (2017)
1	52.1416	38.1899	27.2165	Present
	52.1416	38.1899	27.2165	Zhang et al. (2014)
	52.1416	38.1909	27.2165	Felix et al. (2017)
1/2	65.2867	49.0063	27.2165	Present
1/3	77.5141	59.4309	27.2165	
1/5	105.738	78.2519	27.2165	
1/8	153.629	97.7314	27.2165	
1/12	221.421	111.395	27.2165	

Table 9. Second frequency coefficients of C-C microbeams.

$$\kappa_s = \frac{5}{6}; \nu = 0.38; \frac{L}{h} = 10$$

h/l	MSGT	MCST	CT	
1000	66.8125	66.8123	66.8123	Present
100	66.8316	66.8186	66.8123	
10	68.6635	67.424	66.8123	
5	73.4185	69.109	66.8123	
2	91.6824	78.1659	66.8123	
	91.6824	78.1630	66.8123	Zhang et al. (2014)
1	116.257	97.8845	66.8123	Present
	116.2566	97.8845	66.8123	Zhang et al. (2014)
1/2	165.244	138.616	66.8123	Present
1/3	221.209	171.835	66.8123	
1/5	342.865	211.155	66.8123	
1/8	533.451	234.545	66.8123	
1/12	791.890	245.126	66.8123	

Table 10. Third frequency coefficients of C-C microbeams.

$$\kappa_s = \frac{5}{6}; \nu = 0.38; \frac{L}{h} = 10$$

h/l	MSGT	MCST	CT
1000	116.516	116.516	116.516
100	116.546	116.527	116.516
10	119.409	117.632	116.516
5	126.776	120.765	116.516
2	156.513	138.483	116.516
1	209.665	179.929	116.516
1/2	334.681	257.757	116.516
1/3	472.834	306.509	116.516
1/5	761.424	350.137	116.516
1/8	1203.18	372.24	116.516
1/12	1796.61	382.537	116.516

Tables 8 to 12, contain the first five frequency coefficients of clamped-clamped Timoshenko microbeams. There are three sets of results (MSGT, MCST and CT), carried out by varying thickness-to-material length scale parameter ratio and maintaining constants the values of $\kappa_s = 5/6$, $\nu = 0.38$ and $L = 10h$. When it is possible, comparisons with references (Zhang et al., 2014, Felix et al., 2017) are presented. The calculated results are in excellent agreement with the ones presented in the mentioned references.

Table 11. Fourth frequency coefficients of C-C microbeams.

$$\kappa_s = \frac{5}{6}; \nu = 0.38; \frac{L}{h} = 10$$

h/l	MSGT	MCST	CT
1000	171.899	171.899	171.899
100	171.941	171.917	171.899
10	176.005	173.727	171.899
5	186.566	178.931	171.899
2	233.426	209.141	171.899
1	331.410	279.160	171.899
1/2	563.782	391.532	171.899
1/3	812.975	446.339	171.899
1/5	1325.20	485.764	171.899
1/8	2103.38	501.759	171.899
1/12	3145.90	507.815	171.899

Table 12. Fifth frequency coefficients of C-C microbeams.

$$\kappa_s = \frac{5}{6}; \nu = 0.38; \frac{L}{h} = 10$$

h/l	MSGT	MCST	CT
1000	230.753	230.753	230.776
100	230.812	230.782	230.776
10	236.448	233.552	230.776
5	251.388	241.578	230.776
2	323.648	288.341	230.776
1	485.926	391.848	230.776
1/2	865.869	532.283	230.776
1/3	1266.42	587.419	230.776
1/5	2082.35	626.533	230.776
1/8	3316.00	647.675	230.776
1/12	4965.57	658.613	230.776

Table 13. Frequency coefficients of Clamped- Clamped microbeams. MSGT. $\kappa_s = 5/6; \nu = 0.38; L/h = 20$

h/l	Ω_1	Ω_2	Ω_3	Ω_4	Ω_5
1000	29.6429	78.6932	147.409	231.446	327.313
100	29.6543	78.7218	147.459	231.520	327.411
10	30.7683	81.4995	152.312	238.619	336.843
5	33.8626	89.0971	165.382	257.480	361.627
2	48.9777	123.634	221.414	335.441	462.922
1	73.9684	172.387	297.546	446.707	621.886
1/2	102.603	229.653	414.029	654.618	958.777

Table 14. Fundamental frequency coefficients of C-F microbeams. $\kappa_s = 5/6; \nu = 0.38; L/h = 10$

h/l	**MSGT**	**MCST**	**CT**	
1000	4.74506	4.74504	4.74504	Present
100	4.74695	4.74557	4.74504	
	4.7469	4.7456	4.74504	Felix et al. (2017)
10	4.93176	4.79754	4.74504	Present
5	5.44991	4.94499	4.74504	
	5.4499	4.9451	4.74504	Felix et al. (2017)
2	8.10623	5.78296	4.74504	Present
	8.1062	5.7835	4.74504	Zhang et al. (2014)
	8.1062	5.7836	4.74504	Felix et al. (2017)
1	13.2152	7.50694	4.74504	Present
	13.2152	7.5070	4.74504	Zhang et al. (2014)
	13.2152	7.5072	4.74504	Felix et al. (2017)
1/2	20.8872	9.69872	4.74504	Present
1/3	25.2159	10.6067	4.74504	
1/5	29.1894	13.1165	4.74504	
1/8	31.2555	11.4697	4.74504	
1/12	32.1578	11.5704	4.74504	

Table 13 summarizes the values of frequency coefficients for more slender microbeams, $L = 20\,h$, using the modified strain gradient theory. Once again, values are larger in comparison to the five previous tables with coefficients of shorter beams, $L = 10\,h$.

The following case corresponds to the cantilever microbeam. Calculations are performed adopting clamped and free conditions at the ends of the microbeam, with $\kappa_s = 5/6$, $\nu = 0.38$ and $L = 10\,h$. The results are in Tables 14 to 18. When information is available, these results are compared with the ones from the literature (Zhang et al., 2014, Felix et al., 2017), showing an excellent match.

Table 15. Second frequency coefficients of C-F microbeams.

$$\kappa_s = \frac{5}{6};\ \nu = 0.38;\ \frac{L}{h} = 10$$

h/l	MSGT	MCST	CT	
1000	27.5917	27.5917	27.5916	Present
100	27.6016	27.5946	27.5916	
	27.6016	27.5946	27.5916	Felix et al. (2017)
10	28.5584	27.8808	27.5916	Present
5	31.1646	28.6951	27.5916	
	31.1646	28.6952	27.5916	Felix et al. (2017)
2	42.8270	33.3329	27.5916	Present
	42.8269	33.3346	27.5916	Zhang et al. (2014)
	42.8269	33.3352	27.5916	Felix et al. (2017)
1	59.2227	43.6794	27.5916	Present
	59.2227	43.6792	27.5916	Zhang et al. (2014)
	59.2227	43.6797	27.5916	Felix et al. (2017)
1/2	81.3917	65.0491	27.5916	Present
1/3	102.200	85.3458	27.5916	
1/5	146.389	117.951	27.5916	
1/8	217.152	146.702	27.5916	
1/12	315.326	164.175	27.5916	

Table 16. Third frequency coefficients of C-F microbeams.

$$\kappa_s = 5/6; \nu = 0.38; L/h = 10$$

h/l	MSGT	MCST	CT	
1000	70.0709	70.0708	70.0707	Present
100	70.0932	70.0782	70.0707	
	72.2349	70.8026	70.0707	Felix et al. (2017)
10	70.0932	70.0782	70.0707	Present
5	77.9341	72.8711	70.0707	
	77.9338	72.8767	70.0707	Felix et al. (2017)
2	101.850	84.8665	70.0707	Present
	101.8499	84.8671	70.0707	Felix et al. (2017)
1	138.392	113.13	70.0707	Present
	138.3920	113.1298	70.0707	Felix et al. (2017)
1/2	208.248	170.477	70.0707	Present
1/3	281.972	213.184	70.0707	
1/5	437.018	258.293	70.0707	
1/8	677.768	282.167	70.0707	
1/12	1004.20	292.308	70.0707	

Table 17. Fourth frequency coefficients of C-F microbeams.

$$\kappa_s = 5/6; \nu = 0.38; L/h = 10$$

h/l	MSGT	MCST	CT
1000	122.690	122.690	122.690
100	122.725	122.703	122.690
10	126.112	124.035	122.690
5	134.998	127.873	122.690
2	172.677	150.376	122.690
1	239.571	203.634	122.690
1/2	387.011	298.305	122.690
1/3	546.705	352.925	122.690
1/5	879.133	399.083	122.690
1/8	1387.74	421.809	122.690
1/12	2071.13	432.099	122.690

Table 18. Fifth frequency coefficients of C-F microbeams.

$\kappa_s = 5/6; \nu = 0.38; L/h = 10$

h/l	MSGT	MCST	CT
1000	181.174	181.174	181.174
100	181.210	181.182	181.174
10	185.988	183.329	181.174
5	198.588	189.252	181.174
2	255.393	226.068	181.174
1	370.919	309.421	181.174
1/2	627.471	435.723	181.174
1/3	922.041	492.480	181.174
1/5	1354.01	530.427	181.174
1/8	1973.22	544.696	181.174
1/12	2909.60	549.980	181.174

The frequency coefficients of C-F Timoshenko microbeams with $L = 20\,h$, are included in Table 19 using MSGT.

Other edge conditions were chosen for the following tables. The coefficients of Timoshenko microbeams, clamped at one end and simply supported at the other end (C-S), are presented in Tables 20 to 22.

Table 19. Frequency coefficients of C-F microbeams. MSGT.

$\kappa_s = 5/6; \nu = 0.38; L/h = 20$

h/l	Ω_1	Ω_2	Ω_3	Ω_4	Ω_5
1000	4.79397	29.4412	79.9418	150.315	236.810
100	4.79591	29.4527	79.9718	150.369	236.889
10	4.98571	30.5809	82.8929	155.540	244.537
5	5.52016	33.7308	90.9524	169.616	265.089
2	8.32596	49.5252	129.151	232.729	353.720
1	14.2088	77.8586	189.552	326.560	487.695
1/2	25.5254	116.358	272.816	473.874	733.885

Table 20 shows the non-dimensional frequencies Ω_i, with $i = 1, ..., 5$, applying the modified strain gradient theory with the slenderness parameter $L/h = 10$. Table 21 is similar to Table 20 with the slenderness parameter equal to 20.

Table 22 presents values of non-dimensional frequencies of C-S microbeams calculated with the modified couple stress theory with the slenderness parameter equal to 10. Again, the results are lower than the corresponding MSGT coefficients (Table 20).

Table 20. Frequency coefficients of C-S microbeams. MSGT.

$$\kappa_s = 5/6; \nu = 0.38; L/h = 10$$

h/l	Ω_1		Ω_2		Ω_3		Ω_4	Ω_5
	Present	**Felix et al. (2017)**	**Present**	**Felix et al. (2017)**	**Present**	**Felix et al. (2017)**	**Present**	
1000	19.6733	19.6732	57.8599	57.8597	107.956	107.9557	164.751	225.228
100	19.6804	19.6804	57.8783	57.8783	107.987	107.9867	164.795	225.286
10	20.3748	-	59.6520	-	110.925	-	169.018	231.111
5	22.2735	22.2734	64.3614	64.3599	118.606	118.6039	180.079	246.553
2	30.9051	30.9050	83.8340	83.8325	150.491	150.4895	229.186	320.598
1	43.0923	43.0923	111.645	111.6444	205.976	205.9753	329.274	483.942
1/2	57.4097	-	163.62	-	331.184	-	563.577	810.399

Table 21. Frequency coefficients of C-S microbeams. MCST.

$$\kappa_s = 5/6; \nu = 0.38; L/h = 20$$

h/l	Ω_1	Ω_2	Ω_3	Ω_4	Ω_5
1000	19.6733	57.8598	107.956	164.751	225.226
100	19.6754	57.8659	107.968	164.770	225.258
10	19.8816	58.4622	109.120	166.677	228.154
5	20.4705	60.1722	112.448	172.207	236.579
2	23.7940	69.9467	131.785	204.547	285.555
1	30.9879	92.4862	177.084	278.180	391.722
1/2	44.7353	137.633	257.748	390.703	530.364

Table 22. Frequency coefficients of C-S microbeams. MSGT.

$\kappa_s = 5/6; \nu = 0.38; L/h = 10$

h/l	Ω_1	Ω_2	Ω_3	Ω_4	Ω_5
1000	20.7102	65.1716	130.675	213.136	308.830
100	20.7184	65.1961	130.721	213.207	308.921
10	21.5166	67.5897	135.236	220.104	318.310
5	23.7485	74.2016	147.533	238.633	343.233
2	35.0262	105.680	202.734	318.275	448.051
1	55.7236	155.377	283.512	435.916	613.369
1/2	84.3754	219.788	406.126	649.312	953.759

The next case is the vibrating Timoshenko microbeam with simply supported–free boundary conditions (S-F). Table 23 shows the corresponding frequency coefficients by the modified strain gradient theory (for both $L=10h$ and $L=20h$). As the first mode corresponds to rigid body motion, $\Omega=0$, it has been omitted from the Table.

Table 23. Frequency coefficients of S-F microbeams. MSGT.

$$\kappa_s = \frac{5}{6} \quad \nu = 0.38$$

L/h	h/l	Ω_1	Ω_2	Ω_3	Ω_4	Ω_5
10	1000	20.2871	60.5993	113.873	173.980	238.131
	100	20.2949	60.6206	113.909	174.031	238.193
	10	21.0621	62.6796	117.384	179.002	244.957
	5	23.1967	68.2712	126.645	192.166	262.989
	2	33.7490	93.4022	166.947	251.379	348.753
	1	52.2860	133.007	236.930	368.301	531.347
	1/2	79.7176	203.684	386.633	626.607	672.451
20	1000	20.8825	66.1004	133.178	218.038	318.441
	100	20.8909	66.1260	133.228	218.114	318.555
	10	21.7109	68.6319	138.029	225.532	328.752
	5	24.0147	75.6099	151.239	245.686	356.164
	2	35.9677	110.194	213.293	336.112	475.398
	1	60.0148	171.204	312.671	477.174	667.616
	1/2	101.853	261.581	467.657	728.020	1052.13

Table 24. Frequency coefficients of F-F microbeam. MSGT.

$$\kappa_s = \frac{5}{6} \; \nu = 0.38$$

L/h	h/l	Ω_1	Ω_2	Ω_3	Ω_4	Ω_5
10	1000	29.1533	73.5246	129.589	191.614	257.836
	100	29.1646	73.5504	129.630	191.672	257.916
	10	30.2673	76.0514	133.638	197.269	265.648
	5	33.3362	82.8417	144.329	212.207	286.212
	2	48.5349	113.372	191.251	279.955	383.193
	1	75.5082	161.477	273.803	411.847	587.533
	1/2	115.927	247.377	449.813	646.330	739.669
20	1000	30.2203	81.1484	153.371	243.080	348.630
	100	30.2324	81.1798	153.428	243.165	348.750
	10	31.4182	84.2487	158.946	251.419	360.041
	5	34.7495	92.7899	174.123	273.852	390.334
	2	52.0207	135.016	245.354	374.822	522.578
	1	86.7321	209.179	360.003	534.573	738.422
	1/2	147.439	318.908	541.742	819.687	1169.29

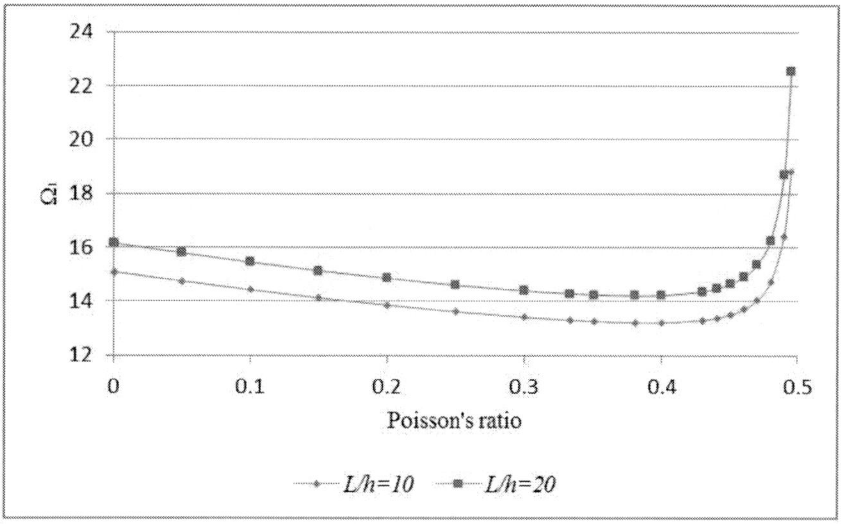

Figure 9. Poisson's ratio effect on fundamental frequency coefficients of C-F microbeams. (MSGT). $\kappa_s = 5/6$; $h = l$.

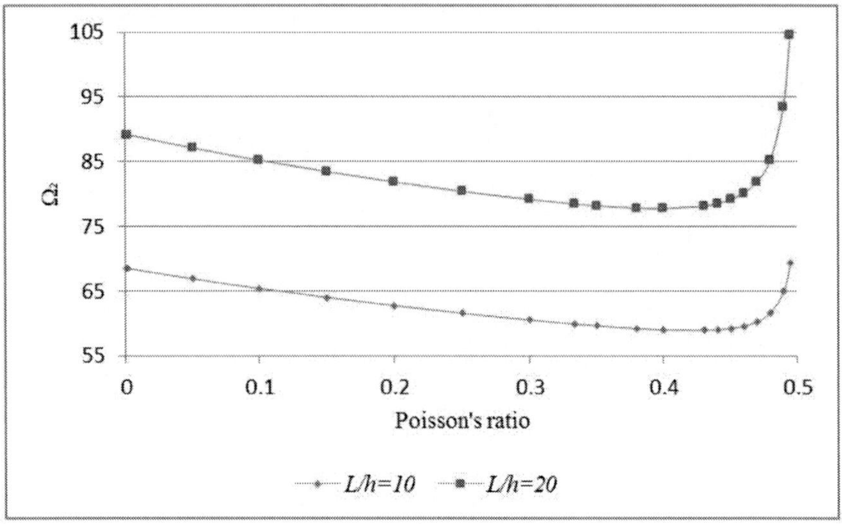

Figure 10. Poisson's ratio effect on second frequency coefficients of C-F microbeams. (MSGT). $\kappa_s = 5/6$; $h = l$.

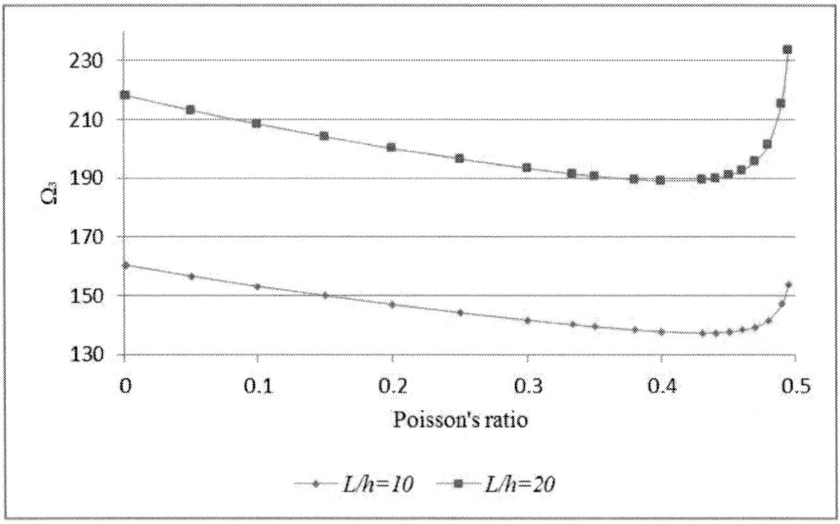

Figure 11. Poisson's ratio effect on third frequency coefficients of C-F microbeams. (MSGT). $\kappa_s = 5/6$; $h = l$.

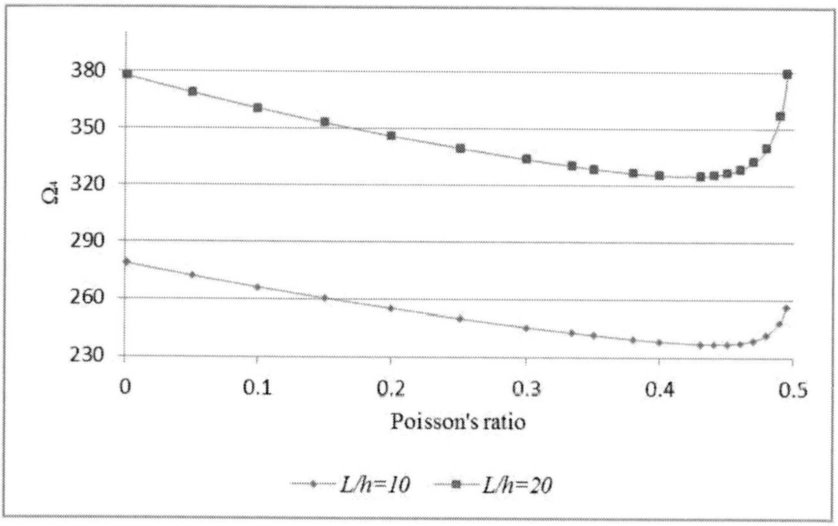

Figure 12. Poisson's ratio effect on fourth frequency coefficients of C-F microbeams. (MSGT). $\kappa_s = 5/6$; $h = l$.

Table 24 contains the values of non-dimensional frequencies of free-free Timoshenko microbeams calculated with the modified strain gradient theory ($L = 10h$ and $L = 20h$). The first and second modes correspond to rigid body movements (translation and rotation), $\Omega = 0$, and have been omitted from Table. Figures 9 to 13 present the behavior of non-dimensional frequencies Ω of Timoshenko microbeams as the Poisson's ratio increases from $(\nu \to 0)$ to $(\nu \to 0.50)$. It is assumed that thickness-to-material length scale parameter ratio is equal to one. It can be observed that results obtained by the modified strain gradient theory have minimum extreme points for the first five frequency coefficients. For the case of the fundamental frequency, the minimum appears when the Poisson's ratio is close to 0.40. For higher frequency coefficients, the minimum extreme points gradually move towards 0.45 (Wang et al., 2010).

Figures 14 to 18 present the first five frequency coefficients of C-C microbeams as the Poisson's ratio increases from $(\nu \to 0)$ to $(\nu \to 0.50)$, with $h / l = 1$. It can be observed a behavior very similar to the behavior in the cantilever beam Figures 9 to 13.

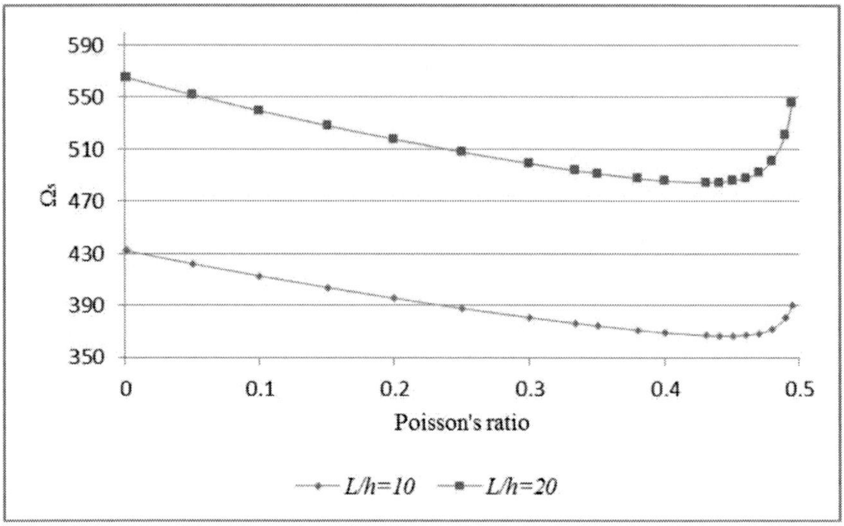

Figure 13. Poisson’s ratio effect on fifth frequency coefficients of C-F microbeams. (MSGT). $\kappa_s = 5/6$; $h = l$.

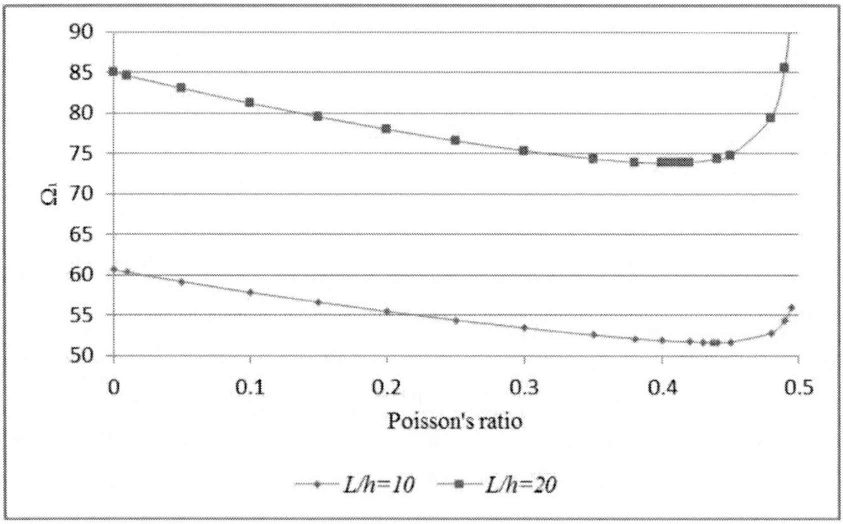

Figure 14. Poisson’s ratio effect on fundamental frequency coefficients of C-C microbeams. (MSGT). $\kappa_s = 5/6$; $h = l$.

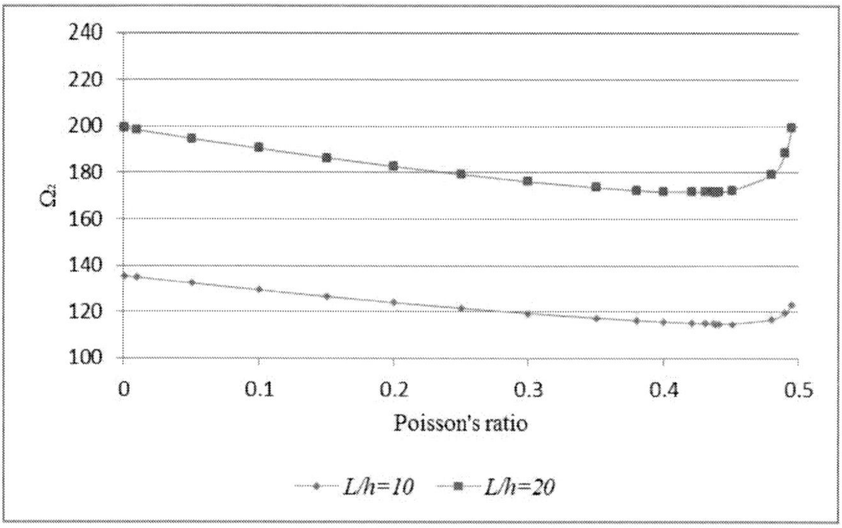

Figure 15. Poisson's ratio effect on second frequency coefficients of C-C microbeams. (MSGT). $\kappa_s = 5/6$; $h = l$.

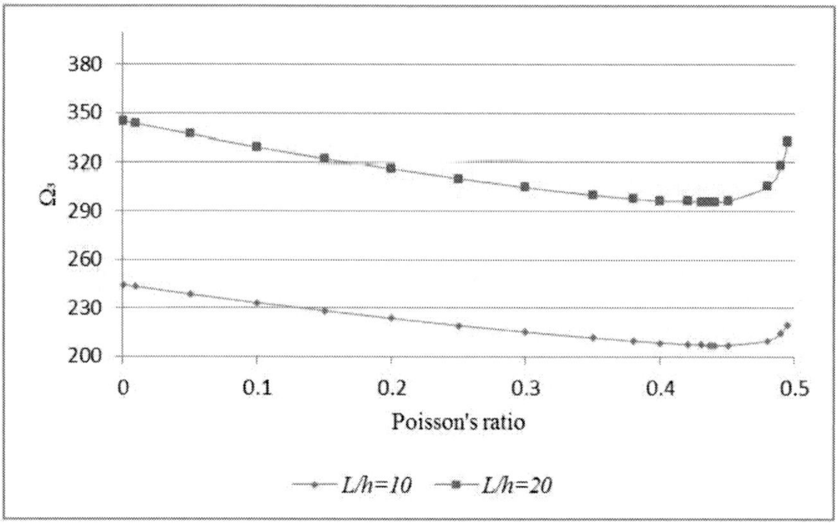

Figure 16. Poisson's ratio effect on third frequency coefficients of C-C microbeams. (MSGT). $\kappa_s = 5/6$; $h = l$.

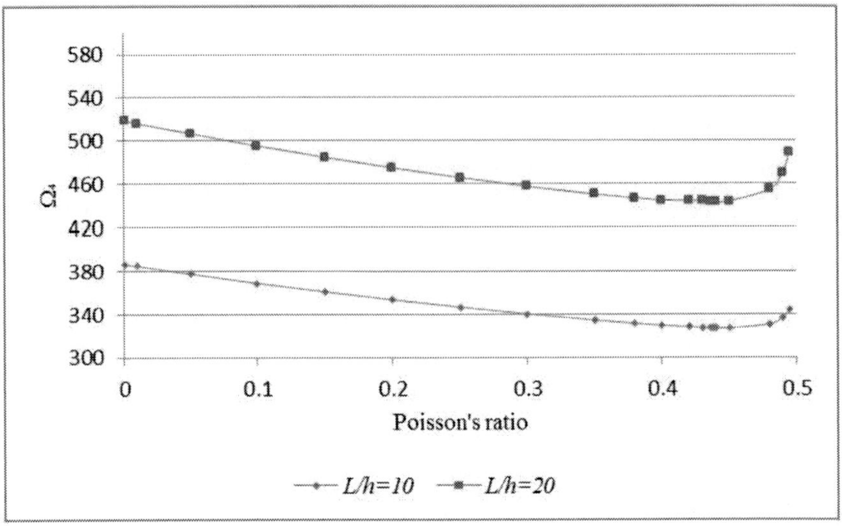

Figure 17. Poisson's ratio effect on fourth frequency coefficients of C-C microbeams. (MSGT). $\kappa_s = 5/6$; $h = l$.

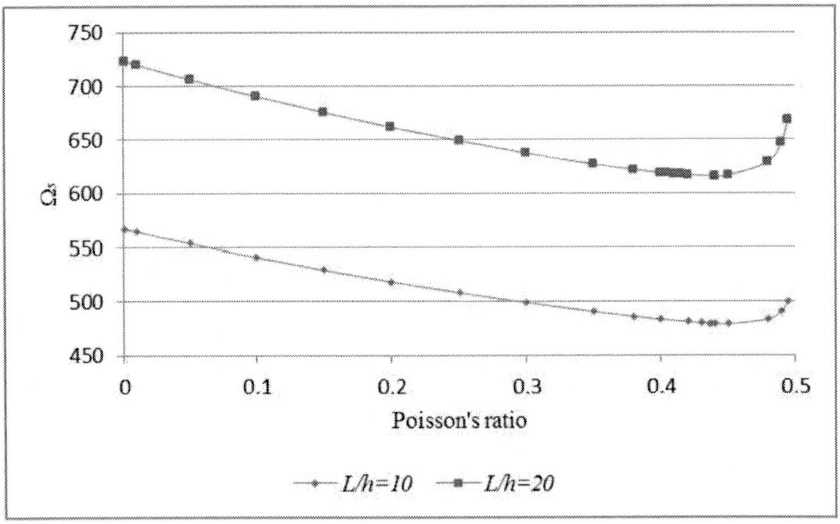

Figure 18. Poisson's ratio effect on fifth frequency coefficients of C-C microbeams. (MSGT). $\kappa_s = 5/6$; $h = l$.

Table 25. Frequency coefficients of S-S microbeams. MSGT.

$$\kappa_s = \frac{5}{6} \quad \nu = 0.38$$

$\frac{L}{h}$	$\frac{h}{l}$	Ω_1	Ω_2	Ω_3	Ω_4	Ω_5	
1000	1000	13.5036	54.0141	121.531	216.125	340.021	
50	100	13.4932	53.7826	120.309	212.454	329.645	
100		13.5052	53.9728	121.260	215.417	336.819	
500		13.5090	54.0341	121.571	216.411	340.404	
1000		13.5091	54.0360	121.581	216.814	339.508	
50	10	14.0291	55.9065	125.016	220.379	342.166	
100		14.0424	56.1165	126.065	223.640	350.061	
500		14.0466	56.1843	126.407	224.390	352.296	
1000		14.0467	56.1864	126.417	224.731	352.462	
B-E		14.0462	56.1785	126.3771			(Akgöz et al., 2013)
50	5	15.5398	61.8838	138.228	243.306	377.056	
100		15.5571	62.1591	139.599	247.539	387.229	
500		15.5627	62.2480	140.048	248.927	390.828	
1000		15.5629	62.2508	140.061	248.842	389.593	
B-E		15.5624	62.2435	140.0248			
B-E		15.56070	62.25337	140.07663			Guerrero et al., 2016)
1000	2	23.5891	94.3542	212.288	377.407	591.124	
50	1	40.5868	158.029	341.388	577.475	857.324	
100		40.8750	162.347	361.110	632.136	973.163	
500		40.9690	163.829	368.439	654.596	1026.68	
1000		40.9719	163.876	368.677	655.393	1028.50	
B-E		40.9724	163.8836	368.7159			(Akgöz et al., 2013)
B-E		40.9708	163.91055	368.85228			(Guerrero et al., 2016)
1000	1/2	78.5301	314.038	706.276	1254.84	1968.14	
B-E		78.53952	314.18779	707.03423			Bernoulli-Euler (Guerrero et al., 2016)

Table 25 shows the non- dimensional natural frequencies predicted by Timoshenko beam theory when the slenderness ratio L / h tends to be very large. As is expected for large slenderness ratios, the present results compare accurately with Bernoulli-Euler results (Akgöz et al., 2013) and (Guerrero et al., 2016).

Obviously, Bernoulli-Euler frequency coefficients are greater than those of short microbeams calculated taking into account effects of shear deformation and rotational inertia. If the relation between length and thickness is large enough, the difference between both beam theories (Timoshenko and Bernoulli-Euler) is small for the fundamental frequency. However, the difference increases for higher frequencies, even for slender microbeams.

CONCLUSION

In this study, a size-dependent Timoshenko beam model in conjunction with modified strain gradient elasticity theory (MSGT) is presented. Micro scale effect, rotatory inertia and shear deformation are taken into consideration. The first five frequency coefficients have been computed using the Ritz method with simply polynomials. Variations of frequency coefficients with the Poisson's ratio are shown in figures.

It is found that

- For micrometric structure dimensions, the effects of the length scale parameters cannot be neglected to obtain results with good precision. Moreover, it can be observed there are significant differences between MSGT, MCST and CT for the analyzed cases of microbeams.
- As expected, the difference in the natural frequency coefficients, calculated with the classical theory (CT) and those obtained with MSGT increases when the thickness-to-material-length scale parameter ratio diminishes.

- In addition, the Poisson's ratio effect is evaluated for frequency coefficients of C-F microbeams and C-C microbeams. Its influence is showed in Figures 9 to 18, where it can be observed that there is minimum extreme point that appears at particular values of Poisson's ratio.
- The results show that the model based on Timoshenko beam theory is particularly important since this beam theory allows describing more accurately both, the vibration of short microbeams and the high modes of vibration of slender and non-slender microbeams.
- In applications where the rotary inertia and the shear deformation effects are not significant, an analysis based on the Bernoulli-Euler beam theory can be applied since Bernoulli-Euler elementary theory only takes into account bending deformation.

ACKNOWLEDGMENTS

The present study has been sponsored by the Universidad Nacional del Sur (UNS), the Consejo Nacional de Investigaciones Científicas y Técnicas (CONICET) and the Comisión de Investigaciones Científicas, Buenos Aires Province (CIC). It was developed in the Department of Engineering and the Engineering Institute (UNS).

REFERENCES

Akgöz, B. & Civalek, Ö. (2012). Analysis of micro-sized beams for various conditions based on the strain gradient elasticity theory. *Archive of Applied Mechanics*, *82*, 423-443.

Akgöz, B. & Civalek, Ö. (2013). A size dependent shear deformation beam model based on the strain gradient elasticity theory. *International Journal of Engineering Science*, *70*, 1-14.

Ansari, R., Gholami, R. & Sahmani, S. (2011). Free Vibration analysis of size-dependent functionally graded microbeams based on strain gradient Timoshenko beam theory. *Composite Structures*, *94*, 221-228.

Araújo Dos Santos, J. V. & Reddy, J. N. (2012). Free vibration and buckling analysis of beams with a modified couple-stress theory. *International Journal of Applied Mechanics*, *4*(3), 125002619.

AZoNano, (2006). *MEMS-Micro Electromechanical Systems and Their Applications Explained and Defined*. [E-book]. https://www.azonano.com/article.aspx?ArticleID=1763.

Bambill, D. V., Guerrero, G. I. & Felix, D. H. (2016). Natural vibrations of micro beams with non-rigid supports. *Journal of Vibration and Control*, 1077546315627727, doi: 10.1177/1077546315627727.

Cosserat, E. & Cosserat, F. (1909). Théorie des Corps Deformables. [Deformable Body Theory] *A. Hermann et Fils*, Paris, reissued by Gabay, Paris, 2009.

Felix, D. H., Guerrero, G. I. & Bambill, D. V. (2016). Análisis dinámico de micro-vigas Timoshenko mediante la teoría elástica del primer gradiente de deformación y el método de elementos finitos. *Mecánica Computacional*, *34*, 3553-3569. [Dynamic analysis of Timoshenko micro-beams using the elastic theory of the first deformation gradient and the finite element method. *Computational mechanics*, *34*, 3553-3569].

Ghayesh, M. H., Amabili, M. & Farokhi, H. (2013). Nonlinear Forced Vibrations of a Microbeam Based on the Strain Gradient Elasticity Theory. *International Journal of Engineering Science*, *63*, 52- 60.

Goodier, J. N. & Timoshenko, S. P. (1970). *Theory of elasticity* (Third edition). New York. U.S.: McGraw-Hill Education.

Guerrero, G. I., Bambill, D. V. & Felix, D. H. (2016). Teoría modificada del gradiente aplicada al análisis de vibraciones libres de una microviga Euler Bernoulli con vínculos elásticos. *Mecánica Computacional*, *34*, 45-564. [Modified stress gradient theory applied to the analysis of free vibrations of a Bernoulli-Euler micro beams with elastic supports. *Computational mechanics*, *34*, 545-564].

Ilanko, S. & Monterrubio, L. E. (2014). *The Rayleigh-Ritz Method for Structural Analysis*. London, U.K.: John Wiley & Sons Inc.

Kahrobaiyan, M. H., Rahaeifard, M., Tajalli, S. A. & Ahmadian, M. T. (2012). A strain gradient functionally graded Euler Bernoulli beam formulation. *International Journal of Engineering Science*, *52*, 65 76.

Kong, S., Zhou, S., Nie, Z. & Wang, K. (2009). Static and dynamics analysis of micro beams based on strain gradient elasticity theory. *International Journal of Engineering Science*, *47*, 487-498.

Lam, D. C. C., Yang, F., Chong, A. C. M., Wang, J. & Tong, P. (2003). Experiments and theory in strain gradient elasticity. *Journal of the Mechanic and Physics of Solids*, *51*, 1477-1508.

Ma, H., Gao, X. & y Reddy, J. N. (2008). A microstructure-dependent Timoshenko beam model based on a modified couple stress theory. *Journal of the Mechanics and Physics of Solids*, *56*, 3379–3391.

Mindlin, R. D. (1964). *Micro-structure in linear elasticity*, Technical Report of Columbia University, Communicated by R.A. Toupin.

Park, S. K. & Gao, X. L. (2006). Bernoulli-Euler beam model based on a modified couple stress theory. *Journal of Micromechanics and Micro-engineering*, *16*(11), 2355-2359.

Şimşek, M. & Reddy, J. N. (2013). Bending and vibration of functionally graded microbeams using a new higher order beam theory and the modified couple stress theory. *International Journal of Engineering Science*, *64*, 37–53.

Timoshenko, S. P. (1921). On the correction for shear of the differential equation for transverse vibrations of prismatic bars. *Philosophical Magazine*, Series, *6*, 41, 744-746.

Timoshenko, S. P. & Goodier, J. N. (1970). *Theory of Elasticity*, (Third Edition). New York, U.S.: McGraw Hill.

Toupin, R. A. (1962). Elastic materials with couple stress. *Archive for Rational Mechanics and Analysis*, *1*(1), 385-414.

Wang, B., Zhao, J. & Zhou, S. (2010). A micro scale Timoshenko beam model based on strain gradient elasticity theory. *European Journal of Mechanics and Solids*, *29*, 591- 599.

Yang, F., Chong, A. C. M., Lam, D. C. C. & Tong, P. (2002). Couple stress based strain gradient theory for elasticity. *International Journal of Solids and Structures*, *39*, 2731-2743.

Zhang, B., He, Y., Liu, D., Gan, Z. & Shen, L. (2014). Non-classical Timoshenko beam element based on the strain gradient elasticity theory. *Finite elements in analysis and design*, *79*, 22-39.

In: Computational Mechanics (CM)
Editor: Jacob Yuen
ISBN: 978-1-53613-672-2

Chapter 3

A PHYSICALLY INTERPRETABLE NOTATION FOR THE FINITE ELEMENT ANALYSIS OF SOLIDS AND STRUCTURES

***J. E. Abdalla Filho*[1,2] *and J. O. Dow*[3,*]**

[1]Graduate Program in Mechanical Engineering,
Pontifical Catholic University of Paraná, Curitiba, Brazil
[2]Graduate Program in Civil Engineering,
Federal Technological University of Paraná, Curitiba, Brazil
[3]Department of Civil, Environmental and Architectural Engineering,
University of Colorado Boulder, Boulder, CO, US

ABSTRACT

The displacement-based finite element method commonly adopts as approximating functions polynomial expansions whose coefficients are unknown. Such coefficients are carried along in the formulation procedure and their physical meanings, although usually unknown to the developer, permeate the entire formulation of the finite element. The use of incomplete and/or inconsistent polynomials will cause spurious terms

* Corresponding Author Email: joao.abdalla@pucpr.br.

to be part of the physical contents of the strain polynomials coefficients. Such spurious terms will be the sources of modeling errors which are known as membrane and shear locking. The deleterious effects of locking have been well described in the literature and their avoidance is paramount for efficient finite element solutions. In general, locking represents quantitative solution errors, meaning that convergence requires a great level of mesh refinement. Further, qualitative errors may arise, and refinement might not be able to overcome them. This work employs a physically interpretable notation called strain gradient notation to study the sources and effects of locking in plane elasticity elements. Such notation allows for the physical meanings of the polynomials coefficients to be identified early in the formulation procedure. This characteristic provides the developer with a clear understanding of the capabilities of the model, and allows for the sources of locking to be easily and precisely identified and removed. Such procedure is carried out without introducing spurious zero-energy modes. Conventional finite element formulations do not possess such capability. For instance, isoparametric finite elements require the use of reduced-order integration procedures, which are not very effective in certain cases as spurious zero-energy modes are introduced. Strain gradient notation guarantees that the sources of locking can be eliminated a-priori, rendering efficient finite element models.

Keywords: finite element method, strain gradient notation, modeling errors, locking, spurious-zero energy modes, parasitic shear, error identification and elimination

INTRODUCTION

It is standard in finite element formulation procedures the adoption of polynomials with unknown coefficients as approximating functions. Although such coefficients possess physical meanings, these are unknown to the developer, which implies that the modeling capabilities of the element cannot be accessed a-priori. Neither can its modeling deficiencies, which is an important issue that will be discussed extensively here.

Modeling deficiencies have always been present in finite element analysis, not only because finite elements have limited modeling capabilities, but also because often they incorporate spurious terms which are directly associated to the approximating functions adopted. Modeling errors were first identified in the framework of isoparametric plane finite elements for stress analysis by Doherty and co-workers [1]. The authors observed that the bilinear element had a poor behavior in pure bending due to 'parasitic shear' (later and up to date called shear locking). That is, some quantity of shear energy would spuriously occur during pure bending, increasing the stiffness of the element. The remedy was to integrate shear terms in the stiffness matrix using only one integration point located at the center of the element. Selective reduced integration was then first proposed, and the term parasitic shear was introduced into the finite element method vocabulary. In the early seventies, parasitic shear was detected in plate and shell analysis. Specifically, it was observed to occur in the isoparametric quadratic shell element developed by Ahmad and co-workers [2] which was developed by degenerating a three dimensional solid element. As the element avoided the Kirchhoff restrictions to thin shell analysis, it should be able to model thick and thin shells. However, it proved to be inefficient in the analysis of thin structures. Zienkiewicz and co-workers [3] pointed that parasitic shear was the cause for Ahmad's element to be overly stiff. They observed that as the thickness decreased progressively, the element's stiffness increased and results became unreliable. Although they applied selective reduced integration, those authors concluded erroneously that a 2x2 uniform reduced integration for all stiffness terms was adequate. Concurrently, Pawsey and Clough [4] treated the same problem, but they arrived at the conclusion that a selective reduced integration scheme should be employed to solve the problem.

Research work conducted by Hughes and co-workers [5, 6, 7] on isoparametric plate elements shed light over what was then called shear locking as well as over another kind of modeling error called spurious zero-energy mode (rank deficiency). Significant progress was made in developing selective reduced integration procedures for plate analysis. Further, observing that selective reduced integration of the nine-node Lagrangean plate element would introduce spurious zero-energy modes,

Hughes and Cohen [8] devised the 'Heterosis' element. Although this element circumvents the problem, it requires a more complex formulation.

Curved elements are appealing for the analysis of curved beams and shells because they have the ability to represent bending and membrane coupling (when flat elements are employed to represent curved structures, such coupling only appears in the assemblage). However, curved elements tend to be overly stiff unless a high-order in-plane displacement field is employed. When low-order in-plane displacement fields are used, a phenomenon called membrane locking occurs. Membrane locking has been shown to result from the effects of the curvature on the bending stiffness due to contributions from the membrane strains [9, 10]. In other words, undue membrane energy is generated during bending because of the inability of the element to bend without stretching.

The aforementioned works can be considered pioneering in the discovery and treatment of the three modeling errors cited, i.e., shear and membrane locking, and spurious zero-energy modes. Since then much effort has been spent on the development of techniques that can eliminate the sources of those deficiencies. Consequently, other important works could be cited here. They are not because that would require a lengthier discussion, which is not our purpose. Instead, we want to show that the use of a physically interpretable notation allows for the precise identification of the sources of those modeling deficiencies in general finite element analysis of solids and structures. Such physically interpretable notation, developed by John O. Dow and several of his graduate students at the University of Colorado Boulder, has been called strain gradient notation. The reason for the name will be explained later. We will show that strain gradient notation allows the analyst to interpret the modeling capabilities of finite elements physically. Further, that it is possible to identify that modeling deficiencies are caused by spurious terms present in strain polynomials, and that, once identified, spurious terms can be readily eliminated a-priori. That is, elimination of spurious terms is done during the formulation procedure, which has the advantage that no numerical techniques are necessary for that purpose during the numerical analysis procedure.

Strain Gradient Notation

Strain gradient notation is a physically interpretable notation which relates displacements, strains and stresses to the kinematic quantities of the continuum. The relationships between displacements and kinematic quantities are derived through an algebraic procedure which identifies the physical contents of polynomial coefficients used as approximate functions in finite elements (details can be found in Dow's book [11]). Important is to note that the modeling characteristics of the finite element being formulated become apparent to the developer since the early steps of the procedure. This means that it is possible to evaluate the capacity of the element a-priori. Also, spurious terms which are the sources of shear and membrane locking can be precisely identified in the proper strain polynomial expansions. This a-priori modeling error identification allows for the elimination of the spurious terms, which is simply done by removing them from the strain polynomials where they reside. The element is then corrected before code implementation which removes the need for the use of numerical remedies such as reduced-order integration, a procedure which is mandatory in isoparametric finite elements (for details of the isoparametric formulation, the reader is referred to Zienkiewicz and Taylor [12]). Further, the precise elimination of the spurious terms avoids the appearance of spurious zero-energy modes as compatible modes are not inadvertently eliminated. This usually happens in isoparametric elements when some form of reduced-order integration is employed to remove sources of locking. The procedure cannot differentiate a spurious term from a legitimate one, and if such terms are of a given order, they will be removed. The deleterious consequence is rank deficiency, and spurious zero-energy modes arise.

Briefly, strain gradient notation was first employed in the context of the analysis of lattice structures [13, 14]. Next, it was employed in the modeling of plane elasticity problems, which is when its ability for spurious terms identification became apparent [15, 16]. Further, it was extended to the analysis of beams and plates [17, 18, 19, 20] and laminated composites [21, 22, 23, 24, 25, 26, 27]. Qualitative errors in finite element modeling of laminated composites were first observed in [21, 22] due to the use of strain gradient notation. Strain gradient notation has also been used in the reformulation of the finite difference method [28, 29, 30, 31]. This will not be discussed here, but it is well documented in Dow's book [11]. The application of strain gradient notation in pointwise error measures has been discussed in Hamernik's doctoral thesis [32]. The present work deals with showing strain gradient notation and its merits in the light of the formulation and application of plane elasticity elements only.

The change in notation promoted by the strain gradient notation can be responsible for improvements to the finite element method. In the following, we present evidence of that through Eqs. 1 and 2:

$$u(x, y) = a_1 + a_2\, x + a_3\, y + a_4\, xy \tag{1a}$$

$$v(x, y) = b_1 + b_2\, x + b_3\, y + b_4\, xy \tag{1b}$$

$$u(x, y) = (u_{rb})_o + (\varepsilon_x)_o\, x + (\gamma_{xy}/2 - r_{rb})_o\, y + (\varepsilon_{x,y})_o\, xy \tag{2a}$$

$$v(x, y) = (v_{rb})_o + (\gamma_{xy}/2 + r_{rb})_o\, x + (\varepsilon_y)_o\, y + (\varepsilon_{y,x})_o\, xy \tag{2b}$$

Equation 1 consists of the standard form of the displacement interpolation polynomials in the x and y directions for a four-node plane elasticity element. Note that the coefficients of Eq. 1 have no discernible physical meanings. In contrast, the coefficients of the interpolation polynomials contained in Eq. 2 display physical meanings. The coefficients $(u_{rb})_o$, $(v_{rb})_o$ and $(r_{rb})_o$ represent rigid body displacements in the x and y directions and the rigid body rotation around the z axis. The coefficients $(\varepsilon_x)_o$, $(\varepsilon_y)_o$ and $(\gamma_{xy})_o$ represent the three constant strains in the continuum. The remaining two terms $(\varepsilon_{x,y})_o$ and $(\varepsilon_{y,x})_o$ identify derivatives of the normal strains in the x and y directions. These terms represent changes in normal strain quantities that can be easily interpreted as bending strains in the x and y directions, respectively. When higher order polynomials are employed, derivatives of these and other strains appear. Thus, this notation is denoted as strain gradient notation. The subscript o indicates that the quantity is a Taylor series coefficient associated with a local origin.

The use of the physically interpretable notation simplifies the element formulation process, provides insights into the element modeling capabilities, and reduces the number of integrals that must be evaluated. In the case of a six-node triangular element the number of integrals is reduced from 78 when the isoparametric formulation is used to 6 when strain gradient notation-based formulation is used. The reduction for a ten node triangle is from 210 to 15 integrals.

The shear strain expressions produced by the two forms of the interpolation polynomials are developed to further demonstrate the power and usefulness of the physically-based notation:

$$\gamma_{xy}(x, y) = \frac{\partial u}{\partial y} + \frac{\partial v}{\partial x} = (a_3 + b_2) + a_4\, x + b_4\, y \tag{3a}$$

$$\gamma_{xy}(x, y) = \frac{\partial u}{\partial y} + \frac{\partial v}{\partial x} = (\gamma_{xy})_o + (\varepsilon_{x,y})_o\, x + (\varepsilon_{y,x})_o\, y \tag{3b}$$

As can be seen, Eq. 3a represents the shear strain as a complete linear polynomial. Without further analysis, this strain representation might be considered as the correct representation of this strain component. However, the visual inspection of Eq. 3b, which is the same expression written in strain gradient notation, identifies the fact that this representation of the shear strain contains the constant representation of the shear strain and the two bending strains identified above. The two normal strain gradient terms, $(\varepsilon_{x,y})$ and $(\varepsilon_{y,x})$, are recognized as spurious terms which are sources of strain modeling errors. These spurious terms can be called parasitic shear terms because they increase the shear strain energy unduly when bending is modeled, for instance. That is, these terms are the sources of shear locking in the four-node plane element because undue shear strain occurs during bending caused by either one of them.

A very good example displaying one of the main features of strain gradient notation has been presented. The reader may at this point realize how powerful the notation can be in formulating other finite elements which are known to behave poorly in given situations. In the remainder of this text, other examples will be discussed to provide a broader insight of the strain gradient notation and modeling errors in finite element analysis. Before that, however, the general formulation procedure will be outlined.

STRAIN GRADIENT-BASED FORMULATION PROCEDURE

The previous section introduces the strain gradient notation and makes it clear that it is a physically-interpretable notation by expressing the polynomial coefficients in terms of kinematic quantities of the continuum. Also, that section shows the power of strain gradient notation in the a-priori evaluation of finite element's modeling capabilities. In turn, this section outlines the general procedure which must be followed to arrive at the stiffness matrix of a finite element departing from the strain gradient approximating functions.

The formulation procedure starts with the proper selection of displacement approximating polynomials in strain gradient notation for a given finite element, which are used to express nodal displacements in terms of strain gradients. The resulting expressions are cast in matrix format as:

$$\{d\} = [\phi]\{\varepsilon_{sg}\} \tag{4}$$

where {d} is the vector of nodal displacements, $[\phi]$ is the matrix which transforms displacements to strain gradients, and $\{\varepsilon_{sg}\}$ is the vector of strain gradients. The strain gradients are the independent deformation modes that the element is capable of representing. Matrix $[\phi]$ is composed of a set of linearly independent vectors, each one associated to one deformation pattern that can be represented by the finite element. That is, each column is associated to one strain gradient. It is easy to see from Eq. (2a) and (2b) that the components ϕ_{ij} are functions of the nodal coordinates.

Strains are defined as the derivatives of the displacement approximating polynomials, and they are represented by the following equation:

$$\{\varepsilon\} = [T_{sg}]\{\varepsilon_{sg}\} \tag{5}$$

where $[T_{sg}]$ is the matrix which transforms strain to strain gradients. It is in this matrix that the spurious terms which cause locking will reside. This knowledge allows the developer to easily remove such spurious terms by eliminating them from $[T_{sg}]$.

The strain energy of the continuum medium can be written as:

$$U = \frac{1}{2}\int_{\Omega} \{\varepsilon\}^T [C]\{\varepsilon\}d\Omega \tag{6}$$

where $[C]$ is the constitutive matrix and Ω is the volume.

Equations (4) and (5) are combined to eliminate $\{\varepsilon_{sg}\}$, and the resulting expression is substituted into Eq. (6) to yield the strain energy in terms of the nodal displacements:

$$U = \frac{1}{2}\{d\}^T[\phi]^{-T} \int_{\Omega} [T_{sg}]^T [C][T_{sg}]d\Omega\, [\phi]^{-1}\{d\} \tag{7}$$

The volume integral contained in Eq. (7) is a matrix that can be represented by $[\bar{U}]$ and is denoted strain gradient matrix. The diagonal terms of this matrix contain the strain energies associated to each deformation mode, and its off-diagonal terms represent the strain energies associated to the various couplings of deformation modes. Evaluation of $[\bar{U}]$ reveals the contribution of each deformation mode and each coupling of deformation modes in terms of strain energy during the deformation of a finite element.

Finally, the finite element stiffness matrix in strain gradient notation is:

$$[K] = [\phi]^{-T}[\bar{U}]\,[\phi]^{-1} \tag{8}$$

The integrals present in $[\bar{U}]$ can be evaluated in closed form, which is an advantage because it removes the need for numerical integration during analysis and the integrated components can be coded easily.

These few steps just outlined have to be followed to formulate the stiffness matrix of any finite element in strain gradient notation. The integrals that have to be carried out in $[\bar{U}]$ are in general very simple and come in a small number, which is very appealing.

Spurious terms appear naturally during the formulation procedure either due to polynomial incompleteness (occurs in various plane elements) or due to inconsistency of polynomial-order with the theory (occurs in beams, plates and shells). As shown in Eq. (3a) and (3b), spurious terms are revealed in the strain polynomial expansions and, thus, appear as components of $[T_{sg}]$. It is part of the strain gradient notation-based formulation procedure to identify and eliminate spurious terms a-priori so that the finite element does not display erroneous behavior. That is, the strain gradient formulation procedure results in elements which are free of modeling errors, and numerical procedures for correction during finite element analysis do not have to be employed.

The next natural step in this chapter is to discuss aspects of the formulation of different finite elements in strain gradient notation. This will be done focusing on modeling characteristics of the elements. Explicit construction of the matrices described above will not be done because it would render a very long text. The references cited in the previous section bring such developments and the reader is referred to them.

STRAIN GRADIENT FINITE ELEMENTS: FORMULATION, ERROR IDENTIFICATION AND ERROR ELIMINATION

This section provides a discussion on the formulation of several strain gradient finite elements for plane analysis. Triangular elements are shown in Figure 1 and quadrilateral elements are shown in Figure 2. The section also shows the modeling capabilities evaluation procedure, which includes the identification and elimination of spurious terms that are introduced during the formulation procedure in each of those elements.

Three-Node Triangle

Solid plane elements used in practice are triangles and quadrilaterals. The three-node triangle, known as CST – Constant Strain Triangle, was introduced by Turner et al. [33] in the inception of the finite element method as an engineering analysis method. This element is shown in Figure 1(a). The approximating functions for this element are given by Eq. (9a) and (9b) in regular notation, and by Eq. (10a) and (10b) in strain gradient notation:

$$u(x, y) = a_1 + a_2 x + a_3 y \tag{9a}$$

$$v(x, y) = b_1 + b_2 x + b_3 y \tag{9b}$$

$$u(x, y) = (u_{rb})_o + (\varepsilon_x)_o x + (\gamma_{xy}/2 - r_{rb})_o y \tag{10a}$$

$$v(x, y) = (v_{rb})_o + (\gamma_{xy}/2 + r_{rb})_o x + (\varepsilon_y)_o y \tag{10b}$$

Consequently, the strain representations in the three-node triangle are:

$$\varepsilon_x = (\varepsilon_x)_o \tag{11a}$$

$$\varepsilon_y = (\varepsilon_y)_o \tag{11b}$$

$$\gamma_{xy} = (\gamma_{xy})_o \tag{11c}$$

which show that the element represents such strains physically correct, although such representations are limited to constant distributions over the domain of the element.

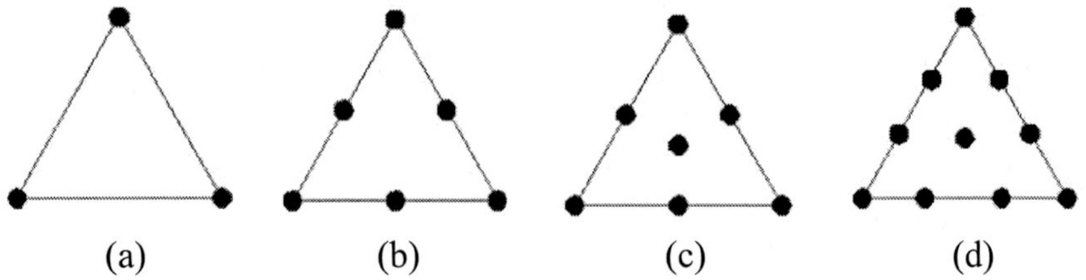

Figure 1. Plane triangular elements: (a) Three-Node, (b) Six-Node, (c) Seven-Node, and (d) Ten-Node.

Six-Node Triangle

When mid-side nodes are inserted in the triangle edges, the six-node triangle is devised (see Figure 1(b)). This element is a higher-order element and it is known as LST – Linear Strain Triangle. Its approximating functions are given in both notations by:

$$u(x, y) = a_1 + a_2\, x + a_3\, y + a_4\, x^2 + a_5\, xy + a_6\, y^2 \tag{12a}$$

$$v(x, y) = b_1 + b_2\, x + b_3\, y + b_4\, x^2 + b_5\, xy + b_6\, y^2 \tag{12b}$$

$$u(x, y) = (u_{rb})_o + (\varepsilon_x)_o\, x + (\gamma_{xy}/2 - r_{rb})_o\, y + (\varepsilon_{x,x}/2)_o\, x^2 + (\varepsilon_{x,y})_o\, xy + [(\gamma_{xy,y} - \varepsilon_{y,x})/2]_o\, y^2 \tag{13a}$$

$$v(x, y) = (v_{rb})_o + (\gamma_{xy/2} + r_{rb})_o\, x + (\varepsilon_y)_o\, y + [(\gamma_{xy,x} - \varepsilon_{x,y})/2]_o\, x^2 + (\varepsilon_{y,x})_o\, xy + (\varepsilon_{y,y}/2)_o\, y^2 \tag{13b}$$

In turn, the strain representations in the six-node triangle are:

$$\varepsilon_x = (\varepsilon_x)_o + (\varepsilon_{x,x})_o\, x + (\varepsilon_{x,y})_o\, y \tag{14a}$$

$$\varepsilon_y = (\varepsilon_y)_o + (\varepsilon_{y,x})_o\, x + (\varepsilon_{y,y})_o\, y \tag{14b}$$

$$\gamma_{xy} = (\gamma_{xy})_o + (\gamma_{xy,x})_o\, x + (\gamma_{xy,y})_o\, y \tag{14c}$$

The displacement polynomials of these two elements show the deformation modes associated to each one of them. It is seen that the CST element contains rigid body motions $(u_{rb})_o$, $(v_{rb})_o$ and $(r_{rb})_o$ and constant strains $(\varepsilon_x)_o$, $(\varepsilon_y)_o$ and $(\gamma_{xy})_o$ in its basis while the LST element also contains first derivatives of all strains. That is, $(\varepsilon_{x,x})_o$, $(\varepsilon_{x,y})_o$, $(\varepsilon_{y,x})_o$, $(\varepsilon_{y,y})_o$, $(\gamma_{xy,x})_o$ and $(\gamma_{xy,y})_o$. Further, analysis of the strain identities of both elements show that all strain components are comprised of terms which are related to them. If one looks closely to Eq. (14a), he or she observes that ε_x is defined by the constant strain $(\varepsilon_x)_o$ and its derivatives $(\varepsilon_{x,x})_o$ and $(\varepsilon_{x,y})_o$. Eq. (14c) shows that the shear strain γ_{xy} is defined by a constant term $(\gamma_{xy})_o$ and its first derivatives $(\gamma_{xy,x})_o$ and $(\gamma_{xy,y})_o$, which are all legitimate terms. The conclusion is that both triangles do not possess parasitic shear terms that would cause increase in strain energy unduly.

Seven-Node Triangle

Figure 1(d) shows the ten-node triangle. If we went ahead on the formulation of the ten-node triangle (this is not done for conciseness), we would observe improvements of the strain expansions and absence of spurious terms. Again, a naturally parasitic shear-free element, which might lead us to the conclusion that triangles are parasitic shear-free elements. However, this would be careless as if we pay attention to the seven-node triangle (see Figure 1(c)) we will notice that it contradicts that belief. As this element has two additional degrees-of-freedom, each of its displacement polynomials will have one extra term. For balance, the u(x,y) expansion will include the term $(\varepsilon_{x,xy}/2)_o\, x^2y$, while the v(x,y) expansion will include the term $(\varepsilon_{y,xy}/2)_o\, xy^2$. The resulting strain expansions are written below:

$$\varepsilon_x = (\varepsilon_x)_o + (\varepsilon_{x,x})_o\, x + (\varepsilon_{x,y})_o\, y + (\varepsilon_{x,xy})_o\, xy \tag{15a}$$

$$\varepsilon_y = (\varepsilon_y)_o + (\varepsilon_{y,x})_o\, x + (\varepsilon_{y,y})_o\, y + (\varepsilon_{y,xy})_o\, xy \tag{15b}$$

$$\gamma_{xy} = (\gamma_{xy})_o + (\gamma_{xy,x})_o\, x + (\gamma_{xy,y})_o\, y + (\varepsilon_{x,xy}/2)_o\, x^2 + (\varepsilon_{y,xy}/2)_o\, y^2 \qquad (15c)$$

Equations (15a) and (15b) show that there are improvements in the representation of normal strains by the seven-node triangle as the quadratic terms $(\varepsilon_{x,xy})_o\, xy$ and $(\varepsilon_{y,xy})_o\, xy$ are added. However, there is no gain in the shear strain representation although two quadratic terms appear in its expansion. Those terms are seen to be second gradients of normal strains, and as they are not genuine terms of the shear strain expansion, it is concluded that they are parasitic shear terms. In a general analysis, such terms cause undue increase of shear strain energy, artificially stiffening the element. That is, those terms are responsible for shear locking. The precise identification of spurious terms, which is peculiar to the strain gradient notation, allows for the complete annihilation of locking if such terms are removed from the shear strain expression. When this is done, Eq. (15c) reduces to Eq. (14c), the shear strain expansion for the six-node triangle, and the seven-node triangle is corrected.

Four-Node Quadrilateral

Next, we change the focus to quadrilateral elements. The four-node plane quadrilateral, which is shown in Figure 2(a), has been used to introduce the strain gradient notation early in this text, and it has been observed that its shear strain polynomial possesses two parasitic shear terms (see Eq. (3b)). After elimination of those two terms, the shear strain of the four-node quadrilateral becomes:

$$\gamma_{xy} = (\gamma_{xy})_o \qquad (16)$$

In the isoparametric formulation, elimination of the effects of parasitic shear terms requires the application of reduced-order integration techniques when calculating stiffness matrices. In the case of the four-node quadrilateral, a one-point Gauss quadrature strategy must be employed.

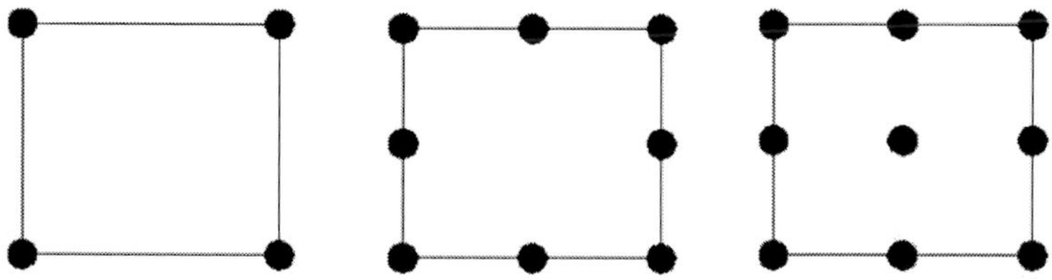

Figure 2. Plane quadrilateral elements: (a) Four-node, (b) Eight-Node, and (c) Nine-Node.

Eight-Node Quadrilateral

The eight-node plane element (see Figure 2(b)), which is obtained by inserting mid-side nodes along the edges of the four-node element and is of the serendipity family, is based on the following strain gradient displacement polynomials:

$$u(x, y) = (u_{rb})_o + (\varepsilon_x)_o\, x + (\gamma_{xy}/2 - r_{rb})_o\, y + (\varepsilon_{x,x}/2)_o\, x^2 + (\varepsilon_{x,y})_o\, xy + [(\gamma_{xy,y} - \varepsilon_{y,x})/2]_o\, y^2 + (\varepsilon_{x,xy}/2)_o\, x^2y + (\varepsilon_{x,yy}/2)_o\, xy^2 \qquad (17a)$$

$$v(x, y) = (v_{rb})_o + (\gamma_{xy/2} + r_{rb})_o\, x + (\varepsilon_y)_o\, y + [(\gamma_{xy,x} - \varepsilon_{x,y})/2]_o\, x^2 + (\varepsilon_{y,x})_o\, xy + (\varepsilon_{y,y}/2)_o\, y^2 + (\varepsilon_{y,xx}/2)_o\, x^2y + (\varepsilon_{y,xy}/2)_o\, xy^2 \qquad (17b)$$

It is observed that four second-order derivatives of normal strains are part of the element deformation basis. These are $(\varepsilon_{x,xy})_o$, $(\varepsilon_{x,yy})_o$, $(\varepsilon_{y,xx})_o$ and $(\varepsilon_{y,xy})_o$, which justify better representations of displacement functions by the eight-node element when compared to the elements described previously. Theoretically, strain modeling by this element should also be superior. In fact, normal strains are incomplete quadratic functions in x and y, which obviously should do a better job than Eq. (14a) and (14b) of the six-node triangle. These polynomials are shown below:

$$\varepsilon_x = (\varepsilon_x)_o + (\varepsilon_{x,x})_o\, x + (\varepsilon_{x,y})_o\, y + (\varepsilon_{x,xy})_o\, xy + (\varepsilon_{x,yy}/2)_o\, y^2 \qquad (18a)$$

$$\varepsilon_y = (\varepsilon_y)_o + (\varepsilon_{y,x})_o\, x + (\varepsilon_{y,y})_o\, y + (\varepsilon_{y,xy})_o\, xy + (\varepsilon_{y,xx}/2)_o\, x^2 \qquad (18b)$$

However, scrutiny of the shear strain polynomial of the eight-node element reveals the presence of spurious terms which will cause deleterious effects in the response of the element. Such polynomial is written below:

$$\gamma_{xy} = (\gamma_{xy})_o + (\gamma_{xy,x})_o\, x + (\gamma_{xy,y})_o\, y + (\varepsilon_{y,xx} + \varepsilon_{x,yy})_o\, xy + 1/2[(\varepsilon_{x,xy})_o\, x^2 + (\varepsilon_{y,xy})_o\, y^2] \qquad (19)$$

Examination of Eq. (19) reveals several features of the shear strain representation of the eight-node element. The first three terms are: a constant shear strain and two first-order derivatives of the shear strain. They are all legitimate terms because they are Taylor series expansion terms of γ_{xy}. The other four remaining terms are second-order derivatives of normal strain components. Following the reasoning applied previously, these four normal strain gradients might be thought of as being parasitic shear terms. The two coefficients of the quadratic terms, $(\varepsilon_{x,xy})_o$ and $(\varepsilon_{y,xy})_o$, are parasitic shear terms which will cause shear locking, and must be removed from Eq. (19). However, it is recognized that the coefficient of the xy term is the following identity:

$$(\gamma_{xy,xy})_o = (\varepsilon_{y,xx} + \varepsilon_{x,yy})_o \qquad (20)$$

which is the compatibility equation of the two-dimensional elasticity. Thus, this is a legitimate term which must be kept in the γ_{xy} strain polynomial. The reader must note that this coefficient could not be recognized as a legitimate term if the expression were not developed in strain gradient notation. After elimination of the spurious terms, the correct shear strain expression for the eight-node quadrilateral is given by:

$$\gamma_{xy} = (\gamma_{xy})_o + (\gamma_{xy,x})_o\, x + (\gamma_{xy,y})_o\, y + (\varepsilon_{y,xx} + \varepsilon_{x,yy})_o\, xy \qquad (21)$$

Elimination of the compatibility equation from Eq. (21) would cause an additional constraint in the shear strain representation because of the removal of the compatibility modes. The result would be another modeling error called spurious zero energy mode.

Nine-Node Quadrilateral

Finally, it is worthwhile to take a quick look into the formulation of the nine-node Lagrange quadrilateral, which is shown in Figure 2(c)). This is done having the eight-node element just discussed as a basis. The element's internal node provides two additional degrees-of-freedom which correspond to the quadratic term x^2y^2 in both u(x, y) and v(x, y) approximating functions. These polynomials are Eq. (17a) and Eq. (17b) plus the terms $(\varepsilon_{x,xyy}/4)_o\, x^2y^2$ and $(\varepsilon_{y,xxy}/4)_o\, x^2y^2$, respectively, which are third-order strain gradients (difficult to interpret their meanings). As a consequence, normal strains are an improvement over Eq. (18a) and Eq. (18b) as the terms $(\varepsilon_{x,xyy}/2)_o\, xy^2$ and $(\varepsilon_{y,xxy}/2)_o\, x^2y$ are added to those polynomials. However, no there is no improvement in the shear strain modeling of the nine-node element. The shear strain polynomial is given by:

$$\gamma_{xy} = (\gamma_{xy})_o + (\gamma_{xy,x})_o\, x + (\gamma_{xy,y})_o\, y + (\varepsilon_{y,xx} + \varepsilon_{x,yy})_o\, xy$$
$$+ 1/2[(\varepsilon_{x,xy})_o\, x^2 + (\varepsilon_{y,xy})_o\, y^2] +$$
$$1/2[(\varepsilon_{x,xyy})_o\, x^2\, y + (\varepsilon_{y,xxy})_o\, xy^2] \qquad (22)$$

Scrutiny of Eq. (22) reveals the presence of the compatibility equation (see Eq. (20)) and two third-order normal strain gradients, which are spurious in the shear strain expression. Elimination of these two terms reduces Eq. (22) to Eq. (21). That is, the eight-node and nine-node quadrilaterals have the same capability in modelling shear strains.

After this discussion on plane elements, one may ask the question why some elements do not incorporate parasitic shear terms and others do. The answer lies on whether the approximating functions are or are not complete polynomials. It is observed that most triangles, for instance, three-node, six-node and ten-node are based on complete polynomials. On the other hand, the quadrilaterals discussed here and the seven-node triangle are based on incomplete polynomials. This is easily seen with the aid of Pascal's triangle [34]. Therefore, the conclusion is that plane elements incorporate parasitic shear terms when their approximating functions are incomplete polynomials.

NUMERICAL ANALYSES AND RESULTS

This section presents numerical analyses of two plane stress problems. The first problem is a long cantilever beam which is modeled using four-node elements, and the second problem is a short cantilever beam which is modeled using eight-node elements. Both problems are analyzed with models with and without parasitic shear terms. The results show the deleterious effects of parasitic shear terms as they cause locking of the models. Elimination of those spurious terms renders locking-free models. Convergence of displacements and stresses described here demonstrate these characteristics.

Long-Beam Problem

The first problem analyzed is shown in Figure 3. It is a clamped-free beam of length L = 1000 mm, height h = 100 mm, and thickness t = 1 mm, subjected to a concentrated load P = 10 kN which is applied at its free end. Material properties of structural steel are employed, namely, Young's modulus E = 200 GPa, Poisson's effect ν = 0.30, and shear modulus G = 87 GPa. Stresses computed at point A and deflections computed at point B are used to show convergence characteristics of the finite element models.

The problem is analyzed with four-node quadrilaterals with and without parasitic shear terms. Four meshes are employed, namely, 4x2 (8 elements), 8x4 (32 elements), 16x4 (64 elements) and 32x8 (256 elements). Figure 4 and Figure 5 show vertical deflections w along the length of the beam when models with parasitic shear (w/PS) and without parasitic shear (wout/PS) are used, respectively.

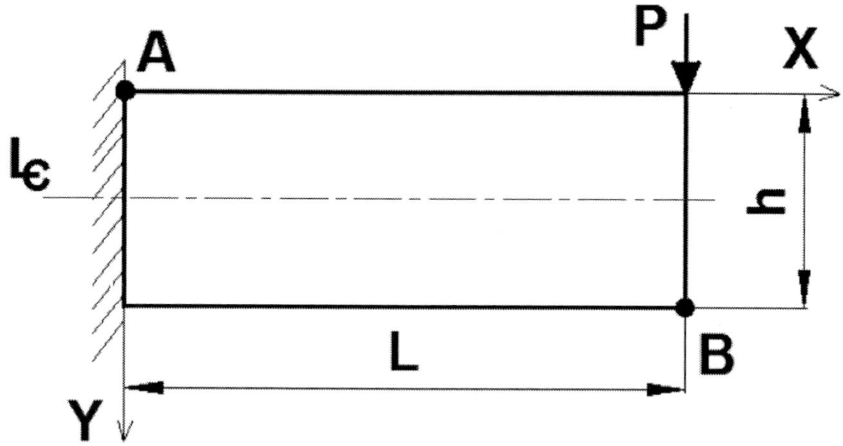

Figure 3. Long clamped-free beam used as example problem.

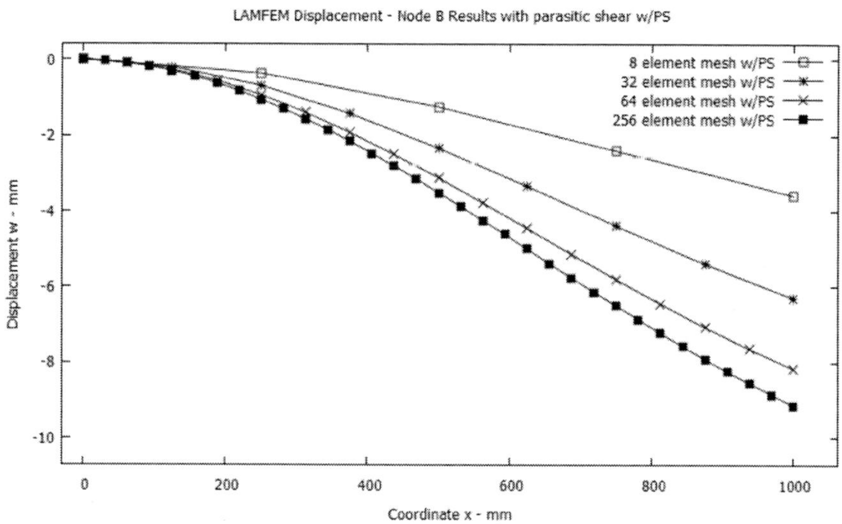

Figure 4. Deflections w at free end. Four-node quadrilateral models with parasitic shear.

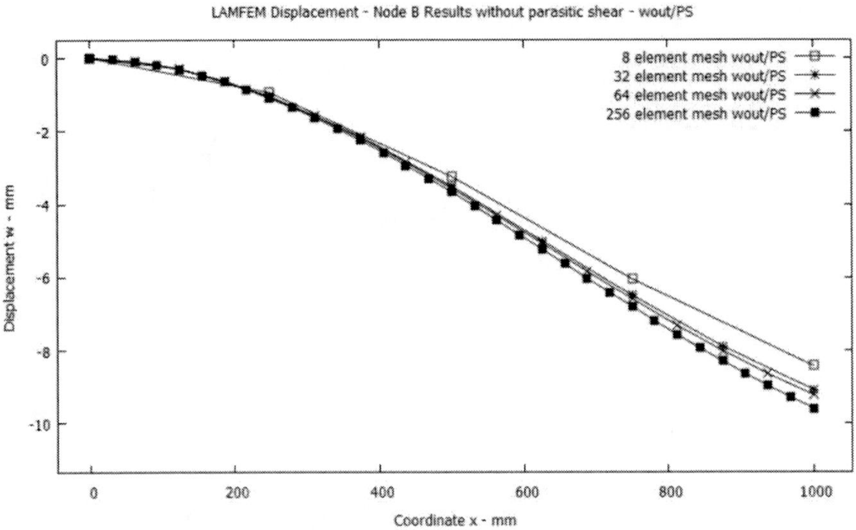

Figure 5. Deflections w at free end. Four-node quadrilateral models without parasitic shear.

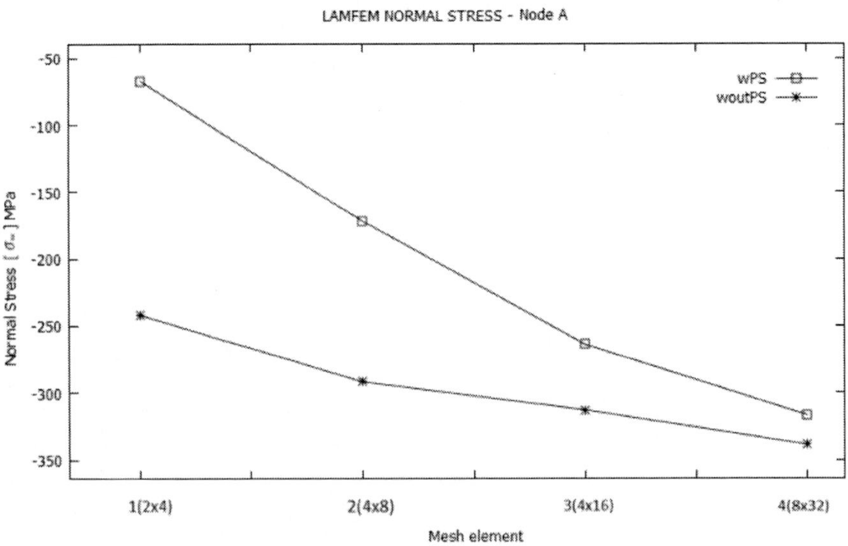

Figure 6. Normal stress at point A. Four-node quadrilateral models.

The better quality of solutions provided by the models without parasitic shear can be seen by comparing both figures. It is evident that convergence occurs much faster for the models without parasitic shear as depicted in Figure 5. The coarser mesh already provides an acceptable solution. The error in the maximum deflection of the coarser mesh with respect to the finer mesh is only 12% for models without parasitic shear. On the other hand, the presence of spurious terms requires a greater level of refinement before the model is capable of providing an acceptable solution. The error in the maximum deflection of the coarser mesh with parasitic shear with respect to the coarser mesh without parasitic shear is nearly 57%. As refinement is performed, such error decreases to nearly 4.8%, reinforcing the idea that shear locking can be attenuated with mesh refinement. Although this is true here in this example problem, a-priori elimination of parasitic shear terms will save computational resources.

Figure 6 and Figure 7 display the normal stress and shear stress values, respectively, calculated at point A shown in Figure 3. Both figures show results computed with models with and without parasitic shear. Figure 6 shows that locking plays an important role in the coarser meshes as results are initially very apart. However, there is the tendency that both solutions tend to converge to the same response of normal stress with refinement. The error in normal stress of the coarser mesh with parasitic shear with respect to the coarser mesh without parasitic shear is nearly 72%, which is unacceptable. Refinement causes the error to decrease to nearly 6.5%, which is a significant reduction.

Figure 7 shows important locking effects caused by the parasitic shear terms on shear stresses. The two curves depicted in the figure show different tendencies. Values of shear stress produced by the model with parasitic shear are much higher than the values produced by the model without parasitic shear, and there is no indication that both models will converge to the same solution. The error in shear stress of the model with parasitic shear with respect to the model without parasitic shear for the finer mesh is nearly 550%, which definitely indicates the need for a-priori elimination of parasitic shear from the model. The reader can find the explanation for the strong erroneous behavior of the model with parasitic shear by reviewing Eq. (3b) for γ_{xy} of the four-node quadrilateral element. As two parasitic shear terms are present in that three-term expression, it is expected that values of shear strain and shear stresses will be very much in error due to the effects of those terms.

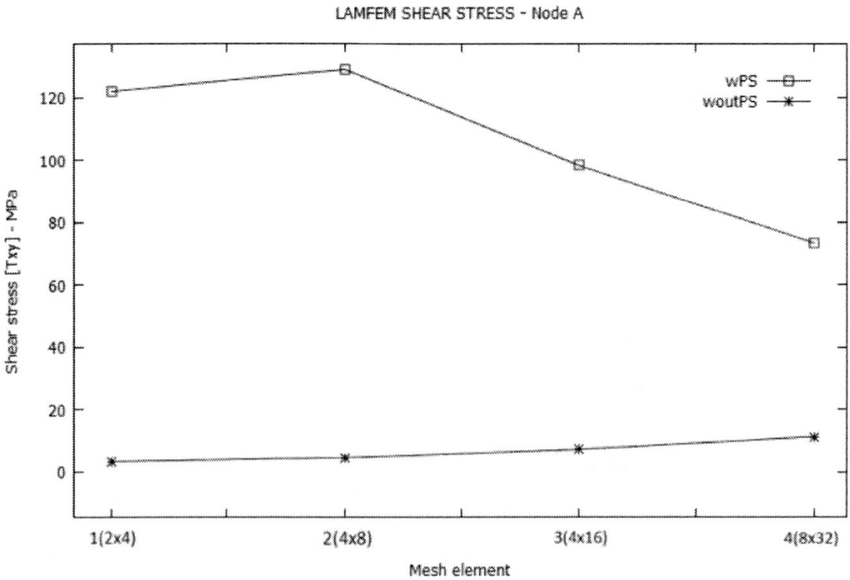

Figure 7. Shear stress at point A. Four-node quadrilateral models.

Short-Beam Problem

The second problem analyzed is shown in Figure 8. It is a short clamped-free beam of length L = 1000 mm, height h = 1000 mm, and thickness t = 20 mm, subjected to a shear load q which is applied at its free end. Material properties of graphite-epoxy are employed. Shear stresses computed at point A are used to show convergence characteristics of the different finite element models.

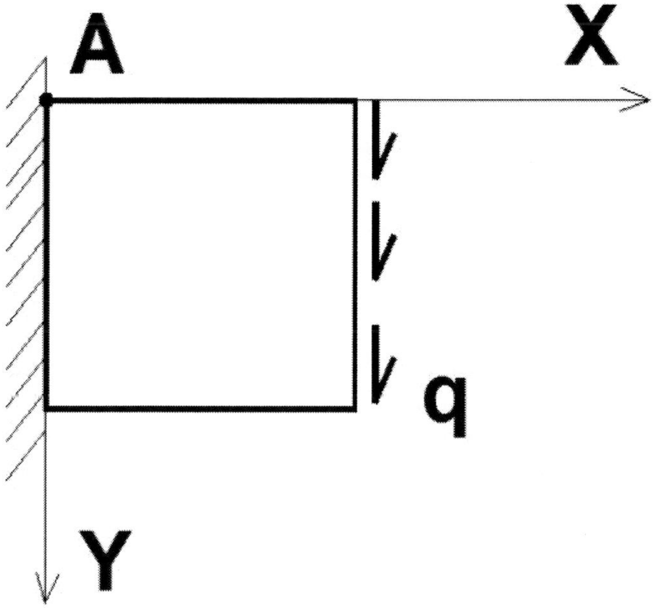

Figurc 8. Short clampcd-frcc bcam uscd as cxamplc problcm.

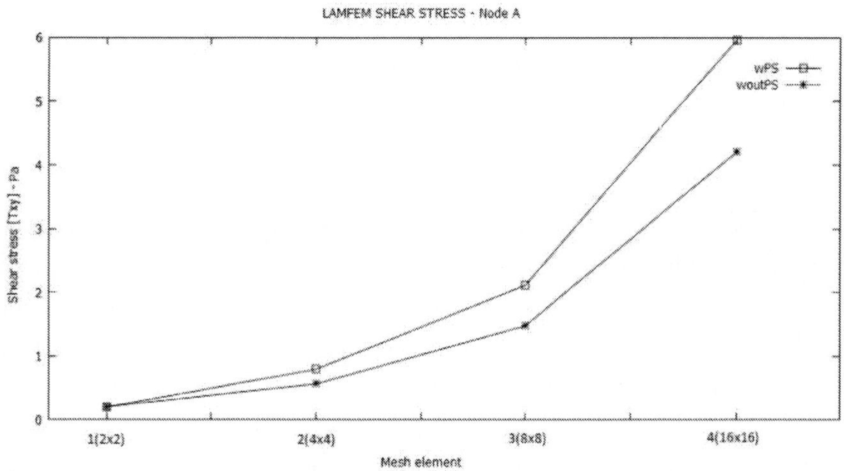

Figure 9. Shear stress at point A. Eight-node quadrilateral models.

The problem is analyzed with eight-node quadrilaterals with and without parasitic shear terms. Four uniform meshes are employed, namely, 2x2 (4 elements), 4x4 (16 elements), 8x8 (64 elements) and 16x16 (256 elements). The results chosen to be discussed here are shear stresses, which are the dominant stresses in a short beam. That is, as bending is nearly inexistent, normal stresses as well as vertical deflections may be neglected. Shear stresses are computed at point A of the clamped end. Figure 9 shows values of shear stress provided by the four models with and without parasitic shear. It is seen that the models with parasitic overestimate the results.

The reader must also observe that the difference between results with and without parasitic shear increases with refinement, which is contrary to what would be expected. For a better flavor of it, percent errors are computed for the four meshes, and they are 2.68%, 28.60%, 30% and 42%.

CONCLUSION

This work presented a physically interpretable notation in the context of the formulation and analysis of the finite element method for solid and structural mechanics called strain gradient notatioin. It has been shown that strain gradient notation allows for the precise identification and elimination of parasitic shear terms, which are responsible for shear locking in plane elasticity elements. Several triangular and quadrilateral elements have been formulated in strain gradient notation, and, whenever present, parasitic shear terms have been identified and then eliminated from the shear strain expansions, which is where they reside.

Two example problems have been analyzed using quadrilateral elements to demonstrate the deleterious effects of shear locking in finite element analysis, and also the importance of correctly removing its sources. It has been shown that the effects of shear locking can be very strong and may not always be removed by mesh refinement. The procedure of removing the spurious terms a-priori is therefore paramount for the quality and efficiency of finite element analysis. Strain gradient notation offers a straightforward procedure for correctly eliminating the sources of shear locking.

ACKNOWLEDGMENTS

The authors would like to thank Tulio Paim Horta, a graduate student at Pontifical Catholic University of Parana, for running the analysis and preparing the figures that have been included in this chapter.

REFERENCES

[1] Doherty, WP; Wilson, EL; Taylor, RL. Stress analysis of axi-symmetric solids using higher-order quadrilateral finite elements, *Structural Engineering Laboratory Report No. SESM 69-3*, University of California, Berkeley, 1969.

[2] Ahmad, S; Irons, BM; Zienkiewicz, OC. Analysis of thick and thin shell structures by curved finite elements, *Int. J. Numer. Meth. Eng.*, 1970, 2, 419-451.

[3] Zienkiewicz, OC; Taylor, RL; Too, JM. Reduced integration technique in general analysis of plates and shells, *Int. J. Numer. Meth. Eng.*, 1971, 3, 275-290.

[4] Pawsey, SF; Clough, RW. Improved numerical integration of thick shell finite elements, *Int. J. Numer. Meth. Eng.*, 1971, 3, 575-586.

[5] Hughes, TJR; Taylor, RL; Kanoknukulchai, W. A simple and efficient finite element for plate bending. *Int. J. Numer. Meth. Eng.*, 1977, 11, 1529-1543.

[6] Hughes, TJR; Cohen, M; Haroun, M. Reduced and selective integration techniques in finite element analysis of plates. *Nucl. Eng. Des.*, 1978, 46, 203-222.

[7] Hughes, TJR; Tezduyar, TE. Finite elements based upon Mindlin plate theory with particular reference to the four-node bilinear isoparametric element. *J. Appl. Mech.*, 1981, 48, 587-596.

[8] Hughes, TJR; Cohen, M. The “heterosis” finite element for plate bending. *Comput. Struct.*, 1978, 9, 445 – 450.

[9] Stolarski, H; Belytschko, T. Membrane locking and reduced integration for curved elements. *J. Appl. Mech.*, 1982, 49, 172 – 176.

[10] Stolarski, H; Belytschko, T. Shear and membrane locking for curved C^o elements. *Comput. Meth. Appl. Mech. Eng.*, 1983, 41, 279 – 296.

[11] Dow, JO. *A Unified Approach to the Finite Element Method and Error Analysis Procedures*; Academic Press: San Diego, CA, 1999.

[12] Zienkiewicz, OC; Taylor, RL. *The Finite Element Method*, Vol. 1, John Wiley & Sons, Inc: New York, NY, 1987.

[13] Dow, JO; Su, ZW; Feng, CC; Bodley, CS. An equivalent continuum representation of structures composed of repeated elements. *AIAA Journal*, 1985, 1564-1569.

[14] Dow, JO; Huyer, SA. Continuum models for space station structures. *ASCE J. Aero. Eng.*, 1989, 212-230.

[15] Dow, JO; Ho, TH; Cabiness, HD. Generalized finite element evaluation procedure. *J. Struct. Eng.*, 1985, 111(2), 435-452.

[16] Dow, JO; Cabiness, HD; Ho, TH. Linear strain element with curved edges. *J. Struct. Eng.*, 1986, 112(4), 692-708.

[17] Byrd, DE. *Identification and Elimination of Errors in Finite Element Analyses*; Doctoral Thesis, University of Colorado Boulder, 1988.

[18] Dow, JO; Byrd, DE. The identification and elimination of artificial stiffening errors in finite elements. *Int. J. Numer. Meth. Eng.*, 1988, 26, 743-762.

[19] Dow, JO; Byrd, DE. Error estimation procedure for plate bending elements. *AIAA Journal*, 1990, 28(4), 685-693.

[20] Mohamed, OA; Byrd, DE; Dow, JO. Improved modeling capabilities with reduced-order integration. *J. Eng. Mech.*, 2005, 131(1), 2-11.

[21] Abdalla Filho, JE. *Qualitative and Discretization Error Analysis of Laminated Composite Plate Models*; Doctoral Thesis, University of Colorado Boulder, 1992.

[22] Dow, JO; Abdalla Filho, JE. Qualitative errors in laminated composite plate models. *Int. J. Numer. Meth. Eng.*, 1994, 37, 1215-1230.

[23] Abdalla Filho, JE; Dow, JO. An error analysis approach for laminated composite plate finite element models. *Comput. Struct.*, 1994, 52(4), 611 – 616.

[24] Abdalla Filho, JE; Fagundes, FA; Machado, RD. Identification and elimination of parasitic shear in a laminated composite beam finite element. *Adv. Eng. Soft.*, 2006, 37, 522-532.

[25] Abdalla Filho, JE; Belo, IM; Pereira, MS. A laminated composite plate finite element a-priori corrected for locking. *Struct. Eng. Mech.*, 2008, 28(5), 603-633.

[26] Abdalla Filho, JE; Belo, IM; Dow, JO. A serendipity plate element free of modeling deficiencies for the analysis of laminated composites. *Compos. Struct.*, 2016, 154, 150-171.

[27] Abdalla Filho, JE; Belo, IM; Dow, JO. On a four-node quadrilateral plate for laminated composites. *Lat. Amer. J. Sol. Struct.*, 2017, 14, 2177-2197.

[28] Dow, JO; Jones, MS; Harwood, SA. A generalized finite difference method for solid mechanics. *Int. J. Numer. Meth. Part. Diff. Eqs.*, 1990, 6 (1), 137-152.

[29] Dow, JO; Jones, MS; Harwood, SA. A new approach to boundary modeling for finite difference applications in solid mechanics. *Int. J. Numer. Meth. Eng.*, 1990, 30 (1), 99-113.

[30] Dow, JO; Stevenson, I. An adaptive refinement procedure for the finite difference method. *Int. J. Numer. Meth. Part. Diff. Eqs.*, 1992, 8 (6), 537-550.

[31] Dow, JO; Hardaway, JL. The modeling of multi-material interfaces in the finite difference method. *Int. J. Numer. Meth. Part. Diff. Eqs.*, 1992, 8 (5), 493-503.

[32] Hamernik, J. A *Unified Approach to Error Analysis in the Finite Element Method*; Doctoral Thesis, University of Colorado Boulder, 1993.

[33] Turner, MJ; Clough, RW; Martin, HC; Topp, LJ. Stiffness and Deflection analysis of Complex Structures, *Journal of Aeronautical Sciences*, 1956, 23, 9, 805-823.

[34] Cook, RD; Malkus, DS; Plesha, ME. *Concepts and Applications of Finite Element Analysis*; Third Edition; John Wiley & Sons, Inc: New York, NY, 1989.

In: Computational Mechanics (CM)
Editor: Jacob Yuen
ISBN: 978-1-53613-672-2

Chapter 4

A Probabilistic Approach for Topology Optimization Analysis Using the FEM

Antônio Rodrigues Neto and Edson Denner Leonel*

University of São Paulo, São Carlos School of Engineering,
São Carlos-SP, Brazil

Abstract

The structural design couples attributes that are, normally, opposites: economy and safety. This type of coupling is handled properly by optimization techniques. Particularly, topology optimization approaches have supported the proposition of structural layouts accounting for the minimum amount of material. Among the topology optimization schemes presented in the literature, it is worth citing the Evolutionary Structural Optimization (ESO). In this approach, the material removal/addition processes are governed by a desired criterion. In the present study, the evolutionary process accounts for threshold values of the Von Mises stress. Therefore, Von Mises stress values smaller than the lower threshold lead to the removal condition. On the other hand, addition

* Corresponding Author Email: edleonel@sc.usp.br.

condition is observed for Von Mises stress values higher than the upper threshold. Moreover, the ESO provides the structural layout for the target volume. Because of the geometry complexity observed during the topology evolution procedure, numerical methods are required for assessing the stress state. In the present study, the Finite Element Method (FEM) is coupled to the ESO for topology optimization analyses. However, this coupling does not account for the inherent randomness that exist into all-mechanical problem. In this regard, this study proposes a numerical scheme for assessing the reliability of structures optimized topologically. The Monte Carlo simulation technique is utilized for determining the probability of structural failure. In addition to the safety aspects, this study clarifies the robustness of the solutions provided by the topology optimization schemes. Two applications are utilized to demonstrate the efficiency of the proposed model.

Keywords: finite element method, topology optimization, structural reliability

INTRODUCTION

The structural layout is proposed for assuring safe and functional conditions taking into account the minimum material consuming. The optimum solution, i.e., the solution that couples perfectly such targets, is not easily available. This solution depends on the material characteristic strengths, structural dimensions and structural system adopted. Neverthless, the optimum layout is achievable through the techniques and concepts proposed in the optimization theory, in which topology optimization theory is a particular case (Bendsøe and Sigmund, 2003). This scientific domain enables the geometrical design of structural components with proper safety accounting for the minimum requirement of material. Thus, this type of optimization achieves the optimum structural geometry, the shape that leads to the minimum requirement of material, respecting constraints related to the stress state at each material point (Chu et al., 1996).

The topology optimization was proposed in the end of 1980s by Bendsøe (1989), which analysed the material distribution into a fixed domain. The optimum structure was achieved through the definition of a material density, which may vary from the void until the full presence of material. The intermediate material density may be also assumed. The bubble method was introduced by Eschenauer et al. (1994). This approach is based on the insertion of cavities into the structure, which enable the optimum solution determination by the shape optimization technique. The simpler Evolutionary Structural Optimization (ESO) method, presented by Xie and Steven (1993), removes material progressively from low stress regions based on a given acceptation/rejection criterion. This method was widely applied with the finite element method (FEM) due to the domain mesh required in both approaches.

This chapter presents a numerical approach for the topology optimization analyses in plane domains. Thus, the proposed numerical model achieves the optimum geometry, i.e., it couples the minimum amount of material to the safety aspects. The optimization technique utilized in this study is the widely known ESO (Querin, 1997, Xie and Steven, 1993). Then, the material removal condition is based on the threshold values for the stress intensity. The Von Mises stress is utilized for defining the global measure of the stress state intensity. In this study, the ESO approach is coupled to the algebraic FEM equations. Therefore, the mechanical fields are determined by the FEM. The linear isoparametric plane FEM elements are utilized for this purpose.

The solutions provided by the topology optimization techniques are optimum from the point of view of material consuming (Sigmund and Petersson, 1998 and Tanskanen, 2002). Thus, such solutions lead to the geometry with the lower material consuming, in which the material points are subjected to the uniform stress state intensity (*fully stressed design approach*). Therefore, the structure topologically optimized has limited capacity for efforts redistribution, because the stress state at its all-material points tend to be uniform and in the neighborhood of the yield or threshold stress. Thus, such structures are more susceptible to the mechanical collapse. Consequently, small material discontinuities such as cracks and/or loading variations may lead to the mechanical collapse. In this context, the reliability analyses have major importance for the topology optimization. This theory enables the proper safety assessment accounting for the failure scenarios and the variables uncertainties.

This study applies the ESO and the FEM for the topology optimization analyses. Two applications are presented, in which the deterministic results are compared with the responses available in the literature. In addition, the robustness analysis is also presented for the optimum structure. In the latter analysis type, the probability of structural failure is assessed by the Monte Carlo simulation method. These analyses demonstrate that, although optimum from the point of view of material consuming, the solution provided by the topology optimization may not be robust when uncertaintines are accounted.

THE FINITE ELEMENT METHOD

The FEM formulation is obtained from the principle of virtual work, which is written as follows:

$$\delta U_e = \delta W_e \tag{1}$$

where δU_e indicates the variation of the virtual internal work and δW_e represents the variation of the virtual external work.

In plane problems, the variables u and v indicate the translations along x and y directions, respectively. These variables compose the vector U, which contains the displacements at the structural plane. Moreover, one defines the vector q as composed of the nodal finite element displacements. Therefore, q contains the displacements u and v of each finite element node. Consequently, the variables U and q can be associated. This relation is performed by applying the shape functions, which approximate the displacements fields along the finite element domain. The f matrix contains the shape functions. Thus:

$$U = fq \tag{2}$$

The strain components are evaluated by differentiating the displacements presented in Eq. (2), as well established into the elasticity theory. Thus, the application of the differential operator *d* leads to the following:

$$\varepsilon = dU \tag{3}$$

Then, the strain components are associated to the nodal displacements by the following equation:

$$\varepsilon = dfq \quad \Rightarrow \quad \varepsilon = Bq \tag{4}$$

where $B = df$. Afterwards, the Hooke's law can be applied, which leads to the following:

$$\sigma = D\varepsilon \quad \Rightarrow \quad \sigma = DBq \tag{5}$$

in which *D* represents the four order tensor composed of elastic properties. Thus, one writes the virtual internal work as follows:

$$\delta U_e = \int_A \delta\varepsilon^T \sigma dA \tag{6}$$

Analogously, the virtual external work assumes the following form

$$\delta W_e = \delta q^T p + \int_A \delta U^T b dA \tag{7}$$

where the vector *p* contains the applied forces and *b* represents the body forces. Therefore, Eq. (1) is rewritten accounting for Eq. (6) and Eq. (7) as follows:

$$\delta U_e = \delta W_e \quad \Rightarrow \quad \int_A \delta\varepsilon^T \sigma dA = \delta q^T p + \int_A \delta U^T b dA \tag{8}$$

The Eq. (8) can be modified taking into account the previous relations introduced in Eq. (4) and Eq. (5). Thus:

$$\int_A B^T DBdA = \delta q^T p + \delta q^T \int_A f^T bdA \quad \Rightarrow \quad \left(\int_A B^T DBdA \right) q = p + \int_A f^T bdA \tag{9}$$

The equation introduced in Eq. (9) is algebraically written as follows:

$$Kq = p + p_b = F \tag{10}$$

where K is named as the stiffness matrix, which is defined by the following expression $K_e = \int_A B^T DBdA$, for each finite element. For plane stress problems, D assumes the following form:

$$D = \frac{E}{1-v^2} \begin{bmatrix} 1 & v & 0 \\ v & 1 & 0 \\ 0 & 0 & \frac{1-v}{2} \end{bmatrix},$$

in which E indicates the Young's modulus and v the Poisson's ratio.

One associates the displacements derivatives to the strains, as well established by the elasticity theory, by the following expressions:

$$\varepsilon_x = \frac{\partial u}{\partial x} \qquad \varepsilon_y = \frac{\partial v}{\partial y} \qquad \gamma_{xy} = \frac{\partial u}{\partial y} + \frac{\partial v}{\partial x} \tag{11}$$

Thus, the operator d is explicitly written as follows:

$$\varepsilon = dU = \begin{bmatrix} \dfrac{\partial u}{\partial x} \\ \dfrac{\partial v}{\partial y} \\ \dfrac{\partial u}{\partial y} + \dfrac{\partial v}{\partial x} \end{bmatrix} \Rightarrow \varepsilon = \begin{bmatrix} \dfrac{\partial}{\partial x} & 0 \\ 0 & \dfrac{\partial}{\partial y} \\ \dfrac{\partial}{\partial y} & \dfrac{\partial}{\partial x} \end{bmatrix} \begin{bmatrix} u \\ v \end{bmatrix} \tag{12}$$

As previously mentioned, the displacements and the element geometry are approximated by the shape functions. For the finite element illustrated in Figure 1, such functions are written as follows:

$$\begin{aligned} \phi_1 &= \frac{(\xi-1)(\eta-1)}{4} & \phi_2 &= \frac{(\xi+1)(1-\eta)}{4} \\ \phi_3 &= \frac{(\xi+1)(\eta+1)}{4} & \phi_4 &= \frac{(1-\xi)(\eta+1)}{4} \end{aligned} \tag{13}$$

Consequently, the Eq. (2) is written explicitly as follows:

$$U = fq \Rightarrow \begin{bmatrix} u \\ v \end{bmatrix} = \begin{bmatrix} \phi_1 & 0 & \phi_2 & 0 & \phi_3 & 0 & \phi_4 & 0 \\ 0 & \phi_1 & 0 & \phi_2 & 0 & \phi_3 & 0 & \phi_4 \end{bmatrix} \begin{bmatrix} u_1 \\ v_1 \\ u_2 \\ v_2 \\ u_3 \\ v_3 \\ u_4 \\ v_4 \end{bmatrix} \tag{14}$$

Because the finite element is described by a dimensionless coordinates ξ and η, the derivatives presented in Eq. (12) are performed by applying the chain rule. Thus, the displacements defined into the Cartesian system are described into the a dimensionless coordinates. In this context, one has:

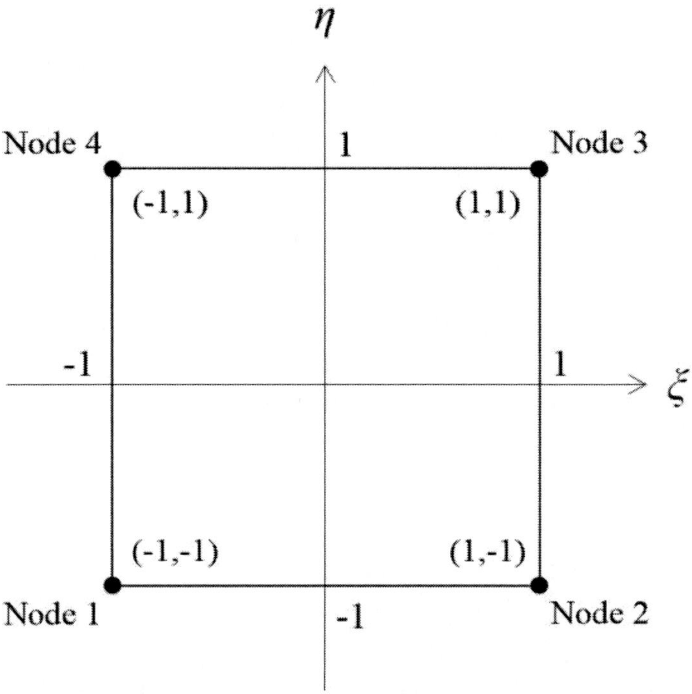

Figure 1. Linear quadrilateral finite element.

$$\frac{\partial \phi}{\partial x} = \frac{\partial \phi}{\partial \xi}\frac{\partial \xi}{\partial x} + \frac{\partial \phi}{\partial \eta}\frac{\partial \eta}{\partial x} \qquad \frac{\partial \phi}{\partial y} = \frac{\partial \phi}{\partial \xi}\frac{\partial \xi}{\partial y} + \frac{\partial \phi}{\partial \eta}\frac{\partial \eta}{\partial y} \tag{15}$$

The last equation is arranged properly as follows:

$$\begin{bmatrix} \frac{\partial \phi}{\partial x} \\ \frac{\partial \phi}{\partial y} \end{bmatrix} = \begin{bmatrix} \frac{\partial \xi}{\partial x} & \frac{\partial \eta}{\partial x} \\ \frac{\partial \xi}{\partial y} & \frac{\partial \eta}{\partial y} \end{bmatrix} \begin{bmatrix} \frac{\partial \phi}{\partial \xi} \\ \frac{\partial \phi}{\partial \eta} \end{bmatrix} \Rightarrow \begin{bmatrix} \frac{\partial \phi}{\partial \xi} \\ \frac{\partial \phi}{\partial \eta} \end{bmatrix} = \begin{bmatrix} \frac{\partial \xi}{\partial x} & \frac{\partial \eta}{\partial x} \\ \frac{\partial \xi}{\partial y} & \frac{\partial \eta}{\partial y} \end{bmatrix}^{-1} \begin{bmatrix} \frac{\partial \phi}{\partial x} \\ \frac{\partial \phi}{\partial y} \end{bmatrix} \tag{16}$$

The global coordinates are associated to the a dimensional coordinates by the Jacobian matrix. This variables is defined as follows:

$$J = \begin{bmatrix} \frac{\partial \phi_1}{\partial \xi} & \frac{\partial \phi_2}{\partial \xi} & \frac{\partial \phi_3}{\partial \xi} & \frac{\partial \phi_4}{\partial \xi} \\ \frac{\partial \phi_1}{\partial \eta} & \frac{\partial \phi_2}{\partial \eta} & \frac{\partial \phi_3}{\partial \eta} & \frac{\partial \phi_4}{\partial \eta} \end{bmatrix} \begin{bmatrix} x_1 & y_1 \\ x_2 & y_2 \\ x_3 & y_3 \\ x_4 & y_4 \end{bmatrix} \Rightarrow J = D_L C_N \tag{17}$$

where C_N contains the global coordinates. By the convenient derivatives of the shape functions, D_L is defined as follows:

$$D_L = \begin{bmatrix} \frac{(\eta-1)}{4} & \frac{(1-\eta)}{4} & \frac{(\eta+1)}{4} & \frac{-(\eta+1)}{4} \\ \frac{(\xi-1)}{4} & \frac{-(\xi+1)}{4} & \frac{(\xi+1)}{4} & \frac{(1-\xi)}{4} \end{bmatrix} \tag{18}$$

Therefore, one associates the local system derivatives to the global system derivatives as follows:

$$D_G = J^{-1} D_L = \begin{bmatrix} \frac{\partial \phi_1}{\partial x} & \frac{\partial \phi_2}{\partial x} & \frac{\partial \phi_3}{\partial x} & \frac{\partial \phi_4}{\partial x} \\ \frac{\partial \phi_1}{\partial y} & \frac{\partial \phi_2}{\partial y} & \frac{\partial \phi_3}{\partial y} & \frac{\partial \phi_4}{\partial y} \end{bmatrix} \tag{19}$$

From the global derivative operator presented in Eq. (19), one defines the operator B. Bearing in mind that $\frac{\partial \phi_i}{\partial x} = D_{G1i}$ and $\frac{\partial \phi_i}{\partial y} = D_{G2i}$ one has:

$$B = \begin{bmatrix} D_{G11} & 0 & D_{G12} & 0 & D_{G13} & 0 & D_{G14} & 0 \\ 0 & D_{G21} & 0 & D_{G22} & 0 & D_{G23} & 0 & D_{G24} \\ D_{G21} & D_{G11} & D_{G22} & D_{G12} & D_{G23} & D_{G13} & D_{G24} & D_{G14} \end{bmatrix} \tag{20}$$

Finally, the differential area dA is defined as follows:

$$dA = \left(\frac{\partial x}{\partial \xi}\frac{\partial y}{\partial \eta} - \frac{\partial y}{\partial \xi}\frac{\partial x}{\partial \eta} \right) d\xi d\eta \;\; \Rightarrow$$

$$dA = \begin{vmatrix} \frac{\partial x}{\partial \xi} & \frac{\partial y}{\partial \xi} \\ \frac{\partial x}{\partial \eta} & \frac{\partial y}{\partial \eta} \end{vmatrix} d\xi d\eta \;\; \Rightarrow \;\; dA = |J| \, d\xi d\eta \tag{21}$$

The stiffness matrix is evaluated utilizing numerical integration technique, in which standard Gauss quadrature with four integration points are applied. It is worth mentioning that only the four integration points enable the exact integration of the kernels. Therefore, the local stiffness matrix is written as follows:

$$K_e = \sum_{i=1}^{4}\sum_{j=1}^{4} B^T D B |J| a_i a_j \tag{22}$$

where a_i and a_j indicate the weights associated to the integration points.

After composing the local stiffness matrix from the all-finite elements, the classical system $KU = F$ is obtained, in which F contains the nodal forces. Thus, this system of algebraic equations is solved by enforcing the boundary conditions at the K matrix, which enable the determination of the nodal displacements.

Thus, after determining the displacements at the mesh nodes, the strain and stress states into each finite element are faster determined as follows:

$$\varepsilon_e = B_e U_e \quad \sigma_e = D_e \varepsilon_e \tag{23}$$

where e indicates the individual finite element.

TOPOLOGY OPTIMIZATION BASED ON THE ESO APPROACH

The topology optimization technique utilized in this study is the Evolutionary Structural Optimization (ESO), which was proposed by Xie and Steven (1993). The kernel of this technique consists on removing material progressively from lower stress intensity regions. The removal criterion is based on the Von Mises stress value. Thus, the finite elements are removed when the Von Mises stress value is smaller than the prescribed threshold. This technique enables the structural layout with maximum stiffness given the target volume. This methodology is known as the ESO based on stress level.

The ESO scheme based on stresses was contested in its origin, because of the lack of mathematical foundations required by the approach. This aspect led to the proposition of one alternative approach: the ESO based on displacements. The ESO displacements approach was based on equalling the total and the strain energies in mechanical problems with stiffness constraint associated to the sensibility number (Chu *et al.*, 1996). Afterwards, the consistency of both ESO methodologies was verified (Zhao *et al.*, 1998). Moreover, Tanskanen (2002) demonstrated that the ESO provides similar responses as those obtained by the non-heuristic approaches such as the Sequential Linear Programming optimization method.

Afterwards the ESO approach consolidation, algorithm variations were proposed to improve the ESO performance. Among such variations, it is worth mentioning: GESO (Genetic Evolutionary Structural Optimization), which integrates operations from genetics (selection, crossing and mutation) to the original ESO; Additive ESO, which is based on the elements addition to the system into regions of high stress concentration; and BESO (Bi-directional Evolutionary Structural Optimization), which adds and removes elements of the system simultaneously (Lanes, 2013). In this chapter, the original ESO method based on stress level is utilized.

THE ESO FORMULATION BASED ON STRESS

The ESO scheme based on stress removes progressively finite elements subjected to the stress level smaller than the prescribed threshold. Thus, mechanically inefficient material is removed and the mechanically efficient design is achieved. Such removal procedure occurs during the evolutionary process, in which at each iteration inefficient material is removed until the target volume is achieved.

The penalty criterion is applied for determining the removal finite elements. This criterion accounts for the Von Mises stress at each element. Thus, the Von Mises stress for each finite element is initially evaluated. Afterwards, these stresses are classified from the highest to the lowest. The finite elements with lower stress intensity, obviously below to the desired analyst threshold, are eliminated. The removal rate, which enables determining the elements for removal condition, is defined as follows:

$$\frac{\sigma_{eq}}{\sigma_{eq\,MAX}} < RR \tag{24}$$

where σ_{eq} represents the Von Mises stress for each finite element. $\sigma_{eq\,MAX}$ indicates the maximum value of the Von Mises stress among the all finite elements into the mesh. The RR parameter is named as the current rejection rate at the given iteration i. Therefore, for each iteration, the finite elements are removed from the mesh if its stress state make true the Eq. (24). Such elements are removed from the mesh in a procedure widely known as hard-kill. Then, the optimization process evolves to the next iteration.

The optimization procedure starts with the prescribed initial rejection rate $RR = RR_i$. However, during the evolutionary process, the Eq. (24) may result false for the all-finite elements. Particularly, this condition is expected to occur during the evolutionary process because the process leads to the stress state constant for the all-finite elements. In this case, non-removal condition is applied. If the target volume has not being achieved, the removal rate is increased. For this case, the evolutionary rate, ER, is utilized for augmenting the removal rate. The removal rate is augmented progressively until the true removal condition for at least one finite element is observed. Thus:

$$RR = RR_i + ER \tag{25}$$

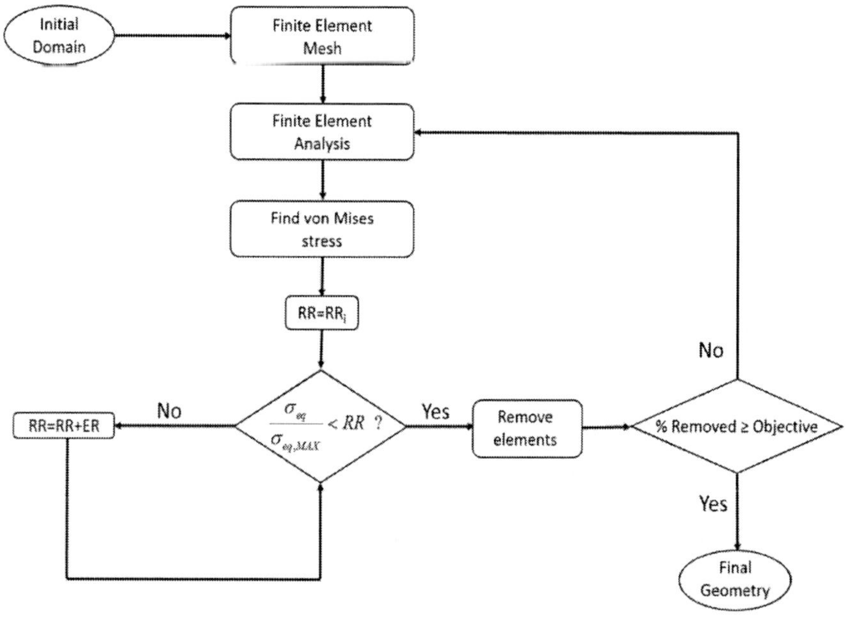

Figure 2. Flowchart for ESO procedures.

It is worth mentioning that RR_i is fixed for the entire evolutionary process. Therefore, the all iterations starts with the same rejection ratio. This strategy leads to the smooth removal procedure. Thus, the optimization algorithm eliminates the finite elements with lower stress level even if the current rejection ratio is smaller than the value utilized in the previous iteration. Querin (1997) suggests small values for the rejection ratio and for the evolutionary rate. The values around 0.01, according to the author, guarantee proper convergence. However, in current structural designs, each domain must be individually analysed and different optimization parameters may be obtained.

The evolutionary process is summarized as follows:

- Step 1: Discretization of the initial domain, construction of the finite element mesh and application of the boundary conditions;
- Step 2: Carry out the finite element analysis; achieve the displacements, stress fields and Von Mises stress values for the all-finite elements; Rating and ranking the Von Mises stress at each finite element;
- Step 3: Removing elements (*hard kill approach*) which satisfy the Eq. (24);
- Step 4: Repeat steps 2 to 4 until the optimum design, i.e., until the target volume.

The Figure 2 presents the flowchart of the entire evolutionary process, as implemented in this study.

STRUCTURAL RELIABILITY

The reliability analysis determines the probability of failure accounting for specific failure scenarios, which is known as the limit states. The probabilistic assessment starts by identifying the random variables involved in the problem $X = [X_1, X_2, ..., X_n]$ and by attributing the specific probability distribution to each random variable (X_i). The probability distribution type assigned for each random variable is defined accounting for physical observations, statistical studies, laboratory analyses and expert opinions. The variables should be assumed as deterministic when they do not interfere significantly into the probability of failure. The probability of failure measures the structural propensity to overcome the limit states.

After the random variables determination, the limit states must be defined. The limit state is the boundary between desired and undesired structural performance. This boundary is often represented by the limit state equation or performance function, G (Nowak and Collins, 2000). Thus, each undesirable structural performance is named as the failure mode. Each failure mode provides one limit state. Failure modes and the corresponding limit states represent ideal modes of failure in structures.

Thus, each limit state equation separates the domain analysis into two regions: the safe domain, where $G_i(X) > 0$ and the failure domain, where $G_i(X) < 0$. The boundary between these two domains is defined as $G_i(x) = 0$, which is named the limit state itself. It is important mentioning that it is not always possible to express, into an explicitly form, the limit states. When numerical methods are utilized for representing the mechanical-material behaviour, solely the limit state function values at a desired amount of points are available. In this study, the limit state function is based on the Von Mises stress value. Therefore, the structural failure is observed when the Von Mises stress value at a given finite element is higher than the yield material stress.

Thus, the probability of failure is calculated as follows:

$$P_f = P\left[\left\{X \in D_f\right\}\right] = P\left[\left\{G_i\left(X\right) < 0\right\}\right] \tag{26}$$

Alternative procedures based on the numerical simulation techniques are often utilized for assessing the probability of failure presented in Eq. (26). In this study, the Monte Carlo simulation technique is utilized for determining the probability of failure accounting for the failure scenario previously presented. The next topic presents detailed description of such method.

THE MONTE CARLO SIMULATION

The Monte Carlo method is a numerical simulation procedure widely utilized for solving reliability problems. In such simulation method, a sampling of random variables is utilized for describing the failure and the safe spaces. The sampling is constructed accounting for the statistical distribution assigned for each random variable in the problem. In iterative problems, such as the addressed in this study where the structural geometry changes at each iteration, the sampling is assessed at each time step. The Monte Carlo approach handles the simulation of the limit state function. Therefore, the spaces' description and the probabilities of failure are accurately achieved as larger be the sampling adopted.

The structural failure is observed when the sampling points leads to the failure domain. Otherwise, safe condition is observed. Optimization approaches, such as the First Order Reliability Method (FORM) and the Second Order Reliability Method (SORM), may be adopted for assessing the probability of failure. These approaches are based on the determination of the reliability index, which is associated to the probability of failure. The probability of failure is calculated, for the Monte Carlo simulation, using the following equation:

$$P_f = \int_{\Omega} I[X] f_x(X) dX = \frac{1}{n_t} \sum_{j=1}^{n} I[X_j] = \frac{n_f}{n_t} \tag{27}$$

in which n_t is the range sampling and n_f the number of failures observed. $I[X]$ assumes 1 for failure condition and nil for safe condition.

It is worth mentioning that Monte Carlo method requires large amount of limit state simulations when the probability of failure is small. This is the disadvantage of this approach. In such case, the computational cost makes this approach non-recommended. Nevertheless, this method is applied in the present study because the finite element model utilized is computationally efficient.

APPLICATIONS

The computational code implemented in this study, which accounts for the probabilistic modelling of topology optimization problems, was utilized for the analysis of two mechanical problems. Such problems were mechanically modelled by the FEM code assuming plane stress state condition. Afterwards, the mechanical responses enabled the topology optimization analysis, in which the optimum structural geometry for specific target volumes is achieved. Finally, the optimum geometries determined by the ESO approach are probabilistically analysed. Different uncertainties scenarios are accounted in order to assess the structural robustness. The deterministic results achieved are in good agreement with the responses available in the literature. Nevertheless, the probabilistic responses illustrate that the solutions provided by the fully-stress design approach are non-robust when randomness is accounted. Therefore, the structural geometries obtained by this approach are more susceptible to the mechanical collapse.

Application 1: Centred Load Case

This application handles the topology optimization analysis for the structure illustrated in Figure 3. This structure consists of a loaded panel, which is fixed at its left end. The loading is composed of a centred force $P = 100$ N applied at the structure right end. The following material properties were adopted in this modelling: Young's modulus $E = 210\,GPa$ and Poisson rate $\upsilon = 0.3$. In this application, the topology optimization procedure aims for determining the optimum geometry for the target volume of 30% of the initial volume.

The mechanical modelling and the topology optimization evolution were performed accounting for three different finite element meshes, which differ from the refinement as illustrated in Figure 4. Such regular meshes have the following dimensions: Mesh 1, with 200 elements of 10 x 10 cm; Mesh 2, with 800 elements of 5 x 5 cm and Mesh 3, with 3200 elements of 2.5 x 2.5 cm. It is worth mentioning that both meshes provided the mechanical convergence. However, one aims to verify the mesh refinement influence over the topology convergence. For the all ESO analysis, one assumed the initial rejection ratio at each iteration equal to $RR_i = 0.02$ and the evolutionary rate equal to $ER_i = 0.01$. These values provide proper convergence and smoothness during the topology evolution.

The Figure 5 presents the geometry evolution process observed during the topology optimization procedure. The all-finite element meshes provided the same optimum structural geometry for the target volume assumed. Nevertheless, it is worth mentioning that the convergence path is different for each initial mesh. Therefore, in complex structural systems, with complex geometry and boundary conditions, the mesh refinement should be verified for the optimum geometry determination. Thus, the convergence analysis is advised in such type of modelling. Moreover, the results achieved for the target volume agree with the responses available in the literature (Vitorio Junior and Leonel, 2017), as illustrated in Figure 6.

The ESO approach leads to the structural geometry in which its all-material points tend to observe the uniform stress state. In this condition, the material characteristic strength is better utilized. Consequently, structures optimized topologically have limited ability for efforts redistribution because its all-material points are subjected to the stress state near to the material strength. The Figure 7 illustrates such condition, in which the Von Mises stress distribution in the beginning and in the end of the analysis are presented. The uniform stress state tendency is observed.

Therefore, the solutions provided by the topology optimization are more susceptible to the mechanical collapse because efforts redistribution is limited. This approach is named as the fully stressed design, which, consequently, provides non-robust solutions when randomness is accounted. The reliability analysis is performed assuming the optimum structure in order to illustrate the non-robustness of the optimum geometry achieved during the topology optimization analysis. The limit state function is written in terms of the Von Mises stress. Then, the stress state caused by the external loading is compared against the material yield stress.

The reliability analysis was performed accounting only for the optimum geometry. In this analysis, only the mesh 2 was considered because such mesh enables the accurate description of the mechanical behaviour. In addition, this mesh provides adequate computational cost. The Monte Carlo simulation method was applied with the sampling size of 100000 for each random variable. The mechanical-material uncertainties were adopted based on Domeneghetti (2011). In addition to the material properties uncertainties, the applied load intensity and its direction were also assumed as random.

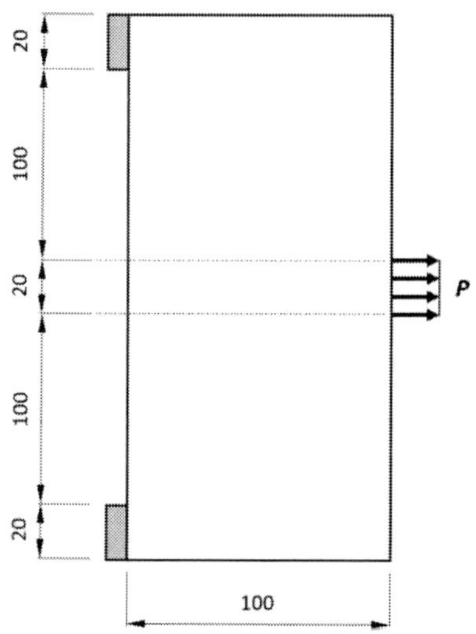

Figure 3. Structure analysed. Dimensions in cm.

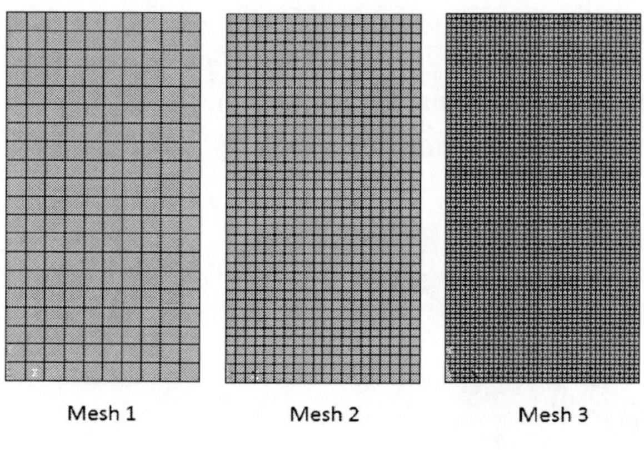

Figure 4. Finite element meshes utilized in this application.

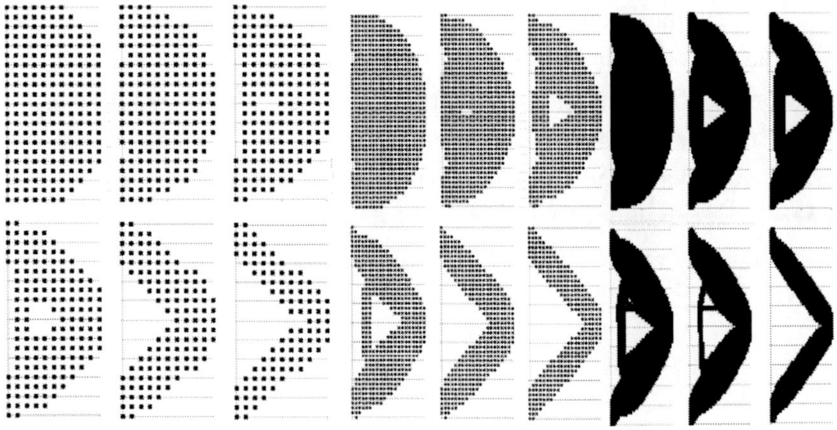

Figure 5. Geometry evolution until 30% of the initial volume for the three meshes utilized.

Four different uncertain scenarios were analyzed, which differ from the amount of random variables and the associated randomness. The Table 1 presents the random variables assumed and the correspondent statistical properties. In addition, the Table 1 also presents the results for the probability of failure assessed on each of such scenarios. One observes that the structural reliability is severely affected by the randomness on the loading orientation. This behaviour is expected because the topology optimization approach proposes the solution accounting for fixed boundary conditions. Thus, the changes on the load intensity and/or its orientation strongly affect the structural reliability. Therefore, the solutions proposed by the ESO based on stresses are not robust for uncertainties. This comment is supported by the behaviour observed in scenarios 3 and 4, in which the probability of failure is strongly increased by the randomness growth of the loading orientation.

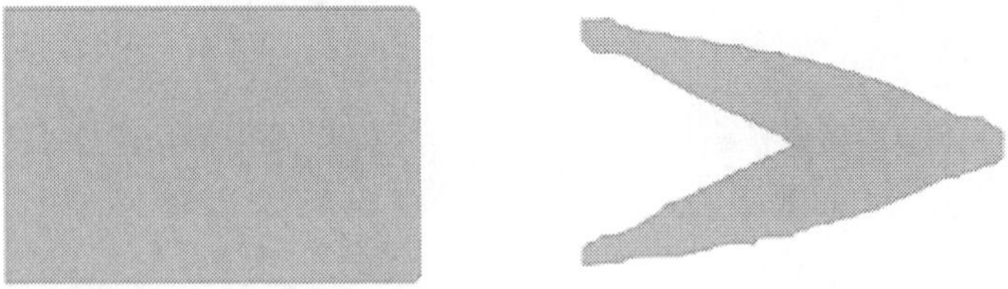

Figure 6. Reference result provided by Vitorio Junior and Leonel (2017).

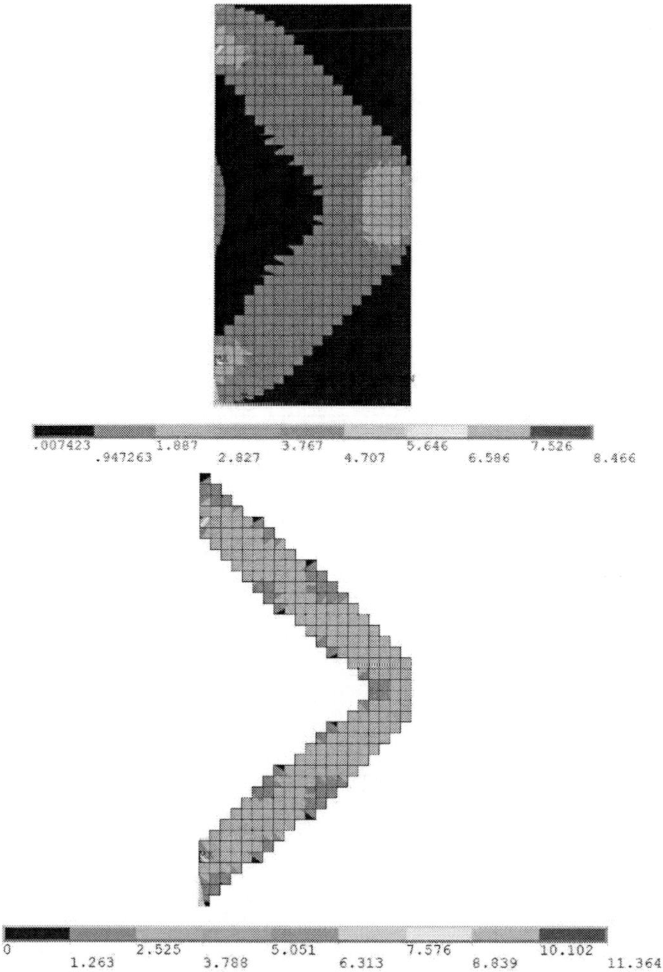

Figure 7. Von Mises stress distribution. Initial and final configurations.

Additional analyses were carried out for verifying the robustness of the solutions provided by the topology optimization scheme for different target volumes. Firstly, the target volumes corresponding to 30%, 50% and 70% of the initial condition were utilized for carrying out the topology optimization procedure. Therefore, after the topology optimization analyses, four different geometries are available, which correspond to the 0%, 30%, 50% and 70% of removal material. For each of these topologically optimized geometries, the reliability analysis was performed assuming the Case 4 described in Table 1. The Figure 8 illustrates the results for these analyses, in which the probability of failure is analysed as a function of the material removal. One observes the high increase on the probability of failure when the structural volume reduces. Thus, such type of analysis may be utilized for defining the optimum target volume as a function of the probability of failure, which leads to the robust solutions.

Table 1. Random parameter for application 1

	Random Variables	**Distribution**	**Mean**	**Standard Deviation**	**Probability of Failure**
Case 1	Yield Stress (kN/cm^2)	Normal	29.8	0.5	2.64%
Case 2	Yield Stress (kN/cm^2)	Normal	29.8	0.5	3.82%
	Nodal Force (kN)	Normal	70	7	
Case 3	Yield Stress (kN/cm^2)	Normal	29.8	0.5	35.35%
	Nodal Force (kN)	Normal	70	7	
	Angle (α°)	Uniform	0	8.66025	
Case 4	Yield Stress (kN/cm^2)	Normal	29.8	0.5	73.16%
	Nodal Force (kN)	Normal	70	7	
	Angle (α°)	Uniform	0	25.98076	

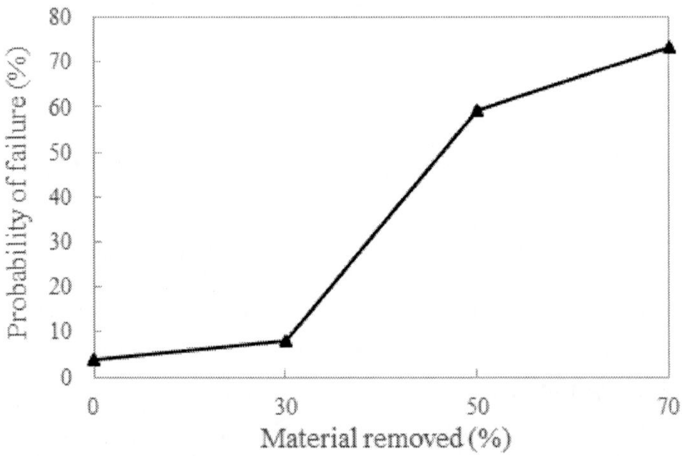

Figure 8. Evaluation of the structural reliability as a function of the removal material.

APPLICATION 2: CENTRED LOAD CASE

The last application of this study concerns the topology optimization analysis of the structure illustrated in Figure 9. This structure is a panel subjected to a concentrated load, which is applied at its right end. This panel is clamped at its left end. The following material properties were assumed in this modelling: Young's modulus $E = 210\,GPa$ and Poisson rate $\upsilon = 0.3$. This application aims for determining the optimum geometry for the target volume of 40% of the initial volume.

The topology optimization analyses were carried out accounting for two different finite element meshes, which differ from the refinement. The following meshes were considered: Mesh 1, with 160 elements of 0.4 x 0.4 cm and Mesh 2, with 640 elements of 0.2 x 0.2 cm. These meshes are illustrated in Figure 10. It is worth mentioning that both meshes enabled the accurate mechanical modelling. However, one verifies the mesh refinement influence during the topology convergence. The following parameters were adopted during the ESO analysis: initial rejection ratio at each iteration equal to $RR_i = 0.02$ and the evolutionary rate equal to $ER_i = 0.01$. These values provide proper convergence and smoothness during the topology evolution.

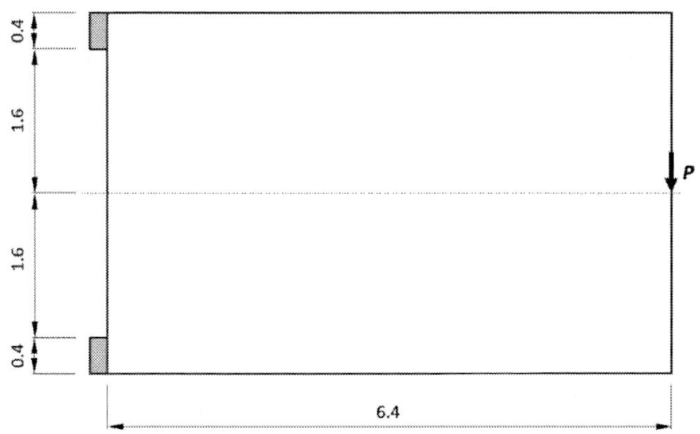

Figure 9. Structure analysed. Dimensions in cm.

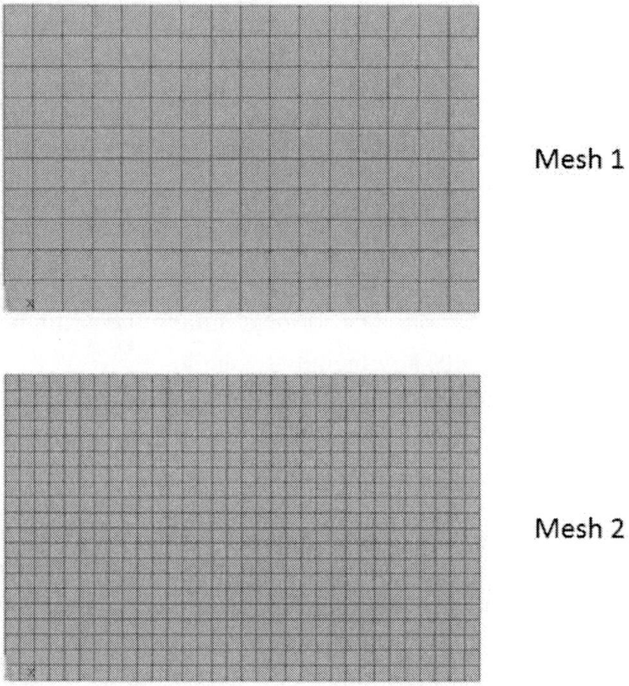

Figure 10. Finite element meshes utilized in this application.

The Figure 11 illustrates the topology optimization convergence for the both finite element meshes adopted. The optimum solutions converge for the same conceptual geometry despite the convergence paths are different. This behaviour indicates that, in general structural systems, the mesh convergence procedure should be accounted. In addition, the solutions achieved agree with the reference response available in the literature (Lanes, 2013), which is illustrated in Figure 12.

The robustness of the topology optimization solutions was also analysed in this application. In this regard, the reliability analyses were performed into the optimum structure geometry to illustrate its non-robustness behaviour when randomness is accounted. The Monte Carlo simulation method was utilized for assessing the probability of failure. The sampling range of 100000 was utilized for each random variable for describing the safe and the failure domains. The random variables and its statistical properties are described in Table 2. In addition, three different uncertain scenarios were considered, which differ from the range of force orientation angle. The reliability analyses were carried out accounting for the mesh 2 because such mesh provided accurate results for the topology procedure and for the mechanical field's description.

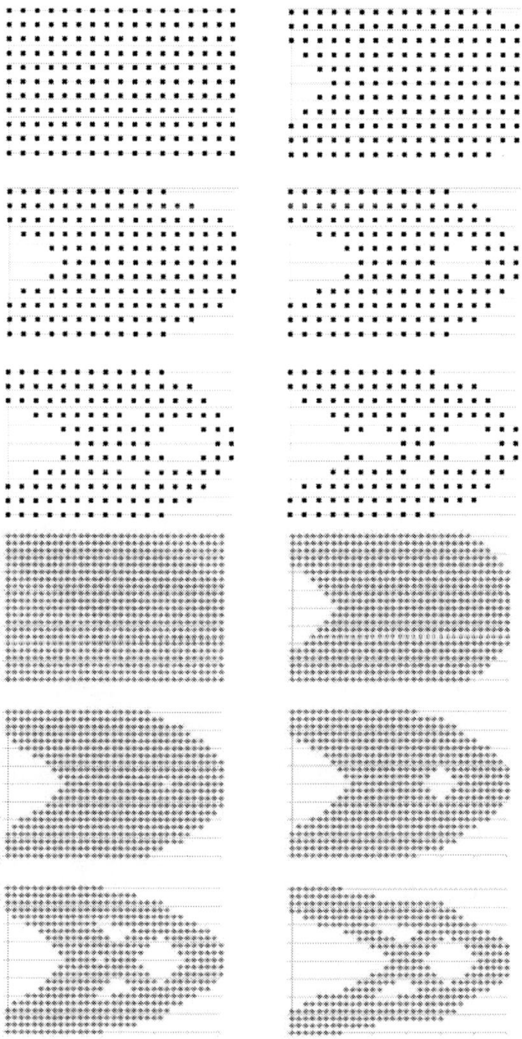

Figure 11. Geometry evolution until 40% of the initial volume for the two meshes utilized.

Figure 12. Reference result provided by Lanes (2013).

The structural reliability is not strongly affected by the loading orientation, as presented in Table 2. Such behaviour is different from the observed in the previous application. In the present case, the increase on the range of the loading orientation leads to the decrease on the probability of failure.

This behaviour of the probability of failure is explained by the influence of the loading effects into the structure. In the first application, the optimum structural geometry was obtained assuming a pure tensile condition. During the probabilistic modelling, the loading orientation change induced bending effects into the structures. Thus, the optimum structural geometry, initially obtained from the pure tensile loading, was subjected to tensile-bending solicitations. The increase on the probability of failure is associated to the stress growth caused by the bending solicitation. On the other hand, the second application was optimized topologically assuming the pure bending condition. Thus, the loading orientation change in such case leads to the tensile-bending solicitation. Therefore, the structure was optimized for the worst condition, which justifies the observed behaviour.

Finally, the reliability analyses were performed for intermediate target volumes. The target volumes of 40%, 60% and 100% of the initial volume were considered. The three different uncertain scenarios presented in Table 2 were utilized. The Figure 13 illustrates the evolution of the probability of failure as a function of the target volume. Similarly to the observed in the previous application, the probability of failure increase as a function of the material removal, as expected. Moreover, the proposed scheme enables determining the optimum target volume as a function of the probability of failure, which introduces the robustness aspects into the solutions.

Table 2. Random parameter for application 2

	Random Variables	Distribution	Mean	Standard Deviation	Probability of Failure
Case 1	Yield Stress (kN/cm^2)	Normal	35.4	1.0	3.25%
	Nodal Force (kN)	Normal	4.0	0.4	
	Angle (α°)	Uniform	0	8.66025	
Case 2	Yield Stress (kN/cm^2)	Normal	35.4	1.0	3.10%
	Nodal Force (kN)	Normal	4.0	0.4	
	Angle (α°)	Uniform	0	17.3205	
Case 3	Yield Stress (kN/cm^2)	Normal	35.4	1.0	2.17%
	Nodal Force (kN)	Normal	4.0	0.4	
	Angle (α°)	Uniform	0	25.98076	

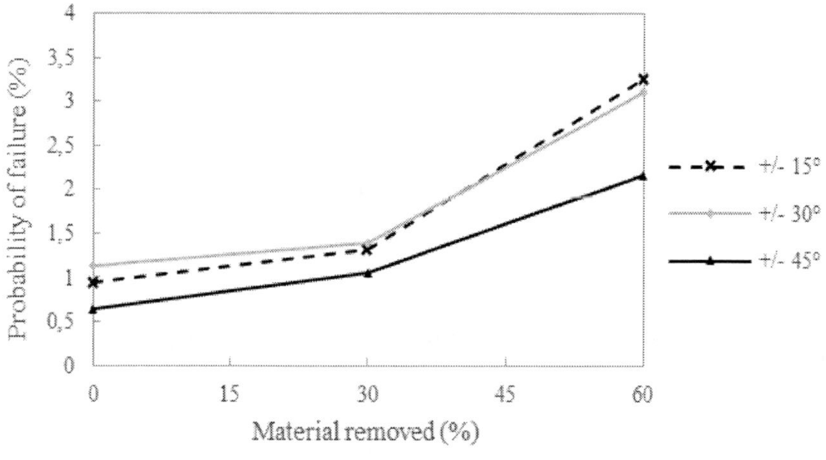

Figure 13. Evaluation of the structural reliability for application 2.

Conclusion

This study presented the ESO approach based on stresses for the topology optimization analysis in plane structural systems. The mechanical structural behaviour is represented by the FEM algebraic equations, which provide the displacements, stresses and strains for each finite element. Two applications were considered for illustrating the efficiency of this approach in obtaining the optimum structural geometries. The ESO approach provided results as accurate as the available in the literature.

However, the responses provided by the topology optimization procedure are not robust for uncertainties. The reliability analyses confirmed that the probability of failure increase as the volume target reduce, as expected. However, for a fixed topology, the optimum solution for instance, the changes on the load direction may affect strongly the structural reliability. Thus, the identification of worst loading conditions have major importance in such type of analysis.

ACKNOWLEDGMENT

The authors would like to thank the sponsorship of this research project by São Paulo Research Foundation (FAPESP), grant number 2014/18928-2.

REFERENCES

Bendsøe, M. P., 1989. “Optimal shape design as a material distribution problem”. *Structural Optimization,* vol. 1, p. 193-202.

Bendsoe, M. P., Sigmund, O., 2003. *Topology optimization: Theory, Methods and Application.* Berlin: Springer-Verlag.

Chu, D. N., Xie, Y. M., Hira, A., Steven, G. P., 1996. “Evolutionary Structural optimization for problems with stiffness constraints”. *Finite Elements in Analysis and Design,* vol. 21, p. 239-251.

Domeneghetti, G, 2011. *A expressão da incerteza de medição em ensaios mecânicos: ISO GUM e Monte Carlo aplicados no ensaio de tração.* Dissertação (Mestrado em Engenharia Mecânica) – Instituto Superior Tupy, Joinville. [*The expression of the measurement uncertainty in mechanical tests: ISO GUM and Monte Carlo applied in the tensile test.* Dissertation (Master in Mechanical Engineering) - Instituto Superior Tupy, Joinville.]

Eschenauer, H., Kobelev, V., Schumacher, A., 1994. "Bubble method for topology and shape optimization of structures". *Struct. Multidiscip. Optim.,* vol. 8, p. 42-51

Lanes, R. M. 2013. *Investigação de um método de otimização topológica evolucionária desenvolvido em script,* Dissertação (Mestrado em Engenharia Civil) – Escola de Engenharia, Universidade Federal de Minas Gerais. [*Investigation of an evolutionary topological optimization method developed in script*, Dissertation (Master in Civil Engineering) - School of Engineering, Federal University of Minas Gerais.]

Nowak, A. S., Collins, K. R., 2000, *Reliability of structures*, Boston: McGraw-Hill.

Querin, O. M., 1997. *Evolutionay structural optimization stress based formulation and implementation.* Tese de Doutorado. Sydney, Australia: University of Sydney.

Sigmund, O., Petersson, J., 1998. "Numerical instabilities in topology optimization: a survey on procedures dealing with checkerboards, mesh dependencies and local minima". *Structural Optimization,* vol. 16, p. 68-75.

Tanskanen, P., 2002 "The evolutionary structural optimization method: theoretical aspects". *Computer Methods in Applied Mechanics and Engineering,* vol. 191, pp. 5485-5498.

Vitorio Junior, P. C., Leonel, E. D., 2017. "Topology optimization analysis based on the direct coupling of the boundary element method and the level set method". *International Journal of Advanced Structural Engineering,* vol. 9, p. 397-407.

Xie, M. Y., Steven, G. P., 1993 "A simple evolutionary procedure for Structural optimization". *Computer & Structures,* vol. 49, p. 885-896.

Zhao, C., Hornby, P., Steven G. P., Xie, Y. M., 1998. "A generalized evolutionary method for numerical topology optimization of structures under static loading conditions". *Structural Optimization,* vol. 15, p. 251-260.

In: Computational Mechanics (CM)
Editor: Jacob Yuen
ISBN: 978-1-53613-672-2

Chapter 5

MATHEMATICAL MODELING OF THE HEATING OF LIQUIDS IN STIRRED TANKS ASSISTED BY MICROWAVES

Javier Ramiro Arballo**, Laura Analía Campañone and Rodolfo Horacio Mascheroni
Center for Research and Development in Food Cryotechnology (CIDCA), National Scientific and Technical Research Council (CONICET), La Plata, Argentina
Department of Chemical Engineering, National University of La Plata, La Plata, Argentina

ABSTRACT

One of the most widely used technologies for food heating today is the use of microwaves. This procedure is applied daily at the domestic level but shows difficulties in industrial designs; one of them is the non-uniformity in the distribution of temperatures. This phenomenon is caused by an inhomogeneous distribution of the electromagnetic fields within the control volume, depending on the dielectric properties of the

* Corresponding Author Email: jrarballo@conicet.gov.ar.

material and the geometry of the system. In this work, the microwave heating of a liquid product inside a tank (reactor) with and without mechanical agitation was analyzed. To this purpose, a mathematical model was developed that includes the resolution of the microscopic energy balance in which a source term is used to consider the interaction between food and microwaves. The electromagnetic distribution was obtained through the resolution of the Maxwell equations for the product and the oven-cavity. In the case of the stirred tank, the microscopic momentum balance was coupled to the non-stationary energy balance to solve the distribution of velocities within the tank. The Maxwell equations and the energy and momentum balances with the initial and boundary conditions were solved using the finite element method through the commercial software COMSOL Multiphysics. From numerical simulations it was possible to evaluate the effect of the agitation rate, the position and number of the microwave-generators inside the cavity in the temperature profiles of the liquid products.

Keywords: heating, microwaves, CFD, stirred tank, liquids

INTRODUCTION

Thermal processing traditional techniques use usually surface heating, where the heat is transferred to the rest of the processed material from a heated surface by conductive mechanism in solids, and by conduction and convection in liquids. Such techniques require the use of hot surfaces, which may result in thermal damage of the product if an increase of processing rates is pursued. Therefore, when these techniques are used, heating rate may be limited by this potential damage on food surface. The volumetric heating (or the simultaneous heating of the entire body of the material) is a technological term describing heating produced throughout the volume of a material. This feature may result in significant benefits in the process and the final quality when applies to food processing.

Volumetric heating techniques are the result of the use of electromagnetic fields at operating frequencies located in various spectrum positions. Mechanisms by which heat is generated in the product vary with the operating frequency of the electric field: the effects in low frequencies (less than 10^4 Hz) and radio frequencies (10^4 to 10^8 Hz) are mainly due to resistive heating caused by the movement of ions/electrons in the electric field. At higher frequencies, for example, microwave heating (10^8 to 10^{11} Hz) the main heating mechanism is dipolar molecular rotation, while at high fixed frequencies (for example, infrared) the dominant mechanism is one of the molecular vibration (Maloney and Harrison, 2016).

Among the techniques, microwaves are widely used. The main use lies in the efficiency improvements due to the reduction of the processing times. Dielectric heating applies to dielectric materials which can interact with radiation. The presence of dissolved salts and the dipolar nature of water molecules are of great importance for dielectric heating. The presence of water in most of the food products allows them to be easily heated in an electromagnetic field (Datta and Anantheswaran, 2005; Arballo et al., 2010; Campañone et al., 2012).

At the domestic level, the use of this technology is of high demand, but the industrial need implies treatment on a larger scale. The irradiation of microwave (MWI) has been used since 1970 as a way of nontraditional heating. After more than 30 years of development, researchers have achieved a good performance on MWI treatment that has been gradually applied from laboratory scale to pilot scale. In chemistry, MW heated reactors are typically used in the laboratory to investigate the viability of manufacturing chemical products. A remarkable feature of batch reactors is the possibility of using directly laboratory research to expand the coverage of the industrial process. Among the advantages of batch reactors is the flexibility of the processes of multiple products, and relatively inexpensive equipment and maintenance cost (Dimian et al., 2014).

Process simulation is a useful tool to study the phenomena brought into play within the equipment. In the heat transfer modeling during microwave heating the interaction of electromagnetic energy with the product should be especially considered. The behavior of electric and magnetic fields within the reactor assisted by microwave is very complex. Maxwell equations describe its distribution within the microwave oven, empty or loaded (Clemens and Saltiel, 1996; Verboven et al., 2003; Dinčov et al., 2004). To model mathematically the process of microwave heating is necessary to formulate and solve the microscopic energy balance, which along with boundary conditions, constitute a system of non-linear differential equations. For the mathematical resolution of heat transfer in transient state there are different numerical methods, which have certain advantages and disadvantages, depending on the characteristics of the problem in study. The most used methods are finite differences (FDM) and finite element (FEM). Although both methods can be used for the space-time discretization, the FEM presents the following benefits compared to the FDM: it applies to irregular geometries, adapts easily to boundary conditions in deformable meshes, not homogeneous food can be analyzed without isotropy and generally with more accuracy (Puri and Anantheswaran, 1993). To facilitate the implementation of the FEM certain commercial softwares have been developed, such as COMSOL Multiphysics. This commercial software has the benefits of having pre-configured modules allowing fast formulation and resolution of the simulations, nevertheless is not easily adaptable to particular conditions of processing. In this work, the software COMSOL Multiphysics was used to analyze the heating of liquid products within the reactors (tanks). For the present analysis, electromagnetic waves and heat transfer in liquids modules were used to obtain the electric field and temperature profiles, respectively. Regarding stirred tanks the laminar flow module was added to incorporate the flow velocity profiles. The aim of this work was to evaluate the effect of the agitation and the number and position of the magnetrons in temperature profiles developed within the liquid sample. Also, obtain an optimal configuration and a proper design considering the lower volumetric variability of temperatures.

Materials and Methods

Mathematical Modeling

A scheme of 3D simulation domains is shown in Figure 1. In the formulation of the complete model, the following considerations for each of the physical mechanisms studied were conducted:

Electromagnetic Field Model

- The absorption of electromagnetic energy by the air was considered neglegible.
- The oven walls were considered perfect conductors.
- The radiation enters at a 2450 MHz frequency in a rectangular waveguide working in a TE_{10} mode.

This formulation considers the energy distribution both in the oven as within the product through the resolution of Maxwell equations:

$$\nabla \times \mu_r^{-1}(\nabla \times E) - k_o^2 \varepsilon_{rc} E = 0 \tag{1}$$

being

$$\varepsilon_{rc} = \varepsilon' - j\varepsilon'' \tag{2}$$

where ε_{rc} is the complex dielectric permittivity, μ_r is the magnetic permeability, ε‘ is the dielectric constant, ε“ is the dielectric loss factor and k_o is the propagation constant defined as:

$$k_o = \alpha + i\beta \tag{3}$$

$$\alpha = \frac{2\pi f}{c}\sqrt{\frac{\varepsilon'(\sqrt{(1+\tan^2\delta)}+1)}{2}} \tag{4}$$

$$\beta = \frac{2\pi f}{c}\sqrt{\frac{\varepsilon'(\sqrt{(1+\tan^2\delta)}-1)}{2}} \tag{5}$$

$$\tan\delta = \frac{\varepsilon''}{\varepsilon'} \tag{6}$$

where β is the attenuation factor, f is the radiation frequency and c is the speed of light.

Equation (1) was solved considering different boundary conditions. For the metal walls of the oven and the waveguide the condition of perfect electric conductor was used:

$$n \times E = 0 \tag{7}$$

The source of electromagnetic energy was incorporated through the boundary condition of the waveguide representing the input port of the microwaves. This rectangular port is excited through a transverse electric wave. Taking into account the operating frequency of the microwave oven (2.45 GHz), the mode TE_{10} was considered as a single propagation mode through the rectangular waveguide.

For food/air interfaces terms of continuity were considered.

Therefore, the absorbed energy per volume unit in this model, Q, due to the interaction of the microwave with food is expressed as:

$$Q(x,y,z,t) = \frac{1}{2}\omega\varepsilon_0\varepsilon''|E|^2 \tag{8}$$

where E is the electric field intensity.

Heat Transfer Model

- Uniform initial temperature within the product,
- Constant thermal, dielectric and transport properties,
- Constant size (without volume changes),
- Convective boundary conditions.

The microscopic energy balance is formulated considering a term of internal generation due to the energy provided by the microwave (Lin et at., 1995). The resulting microscopic balance is as follows:

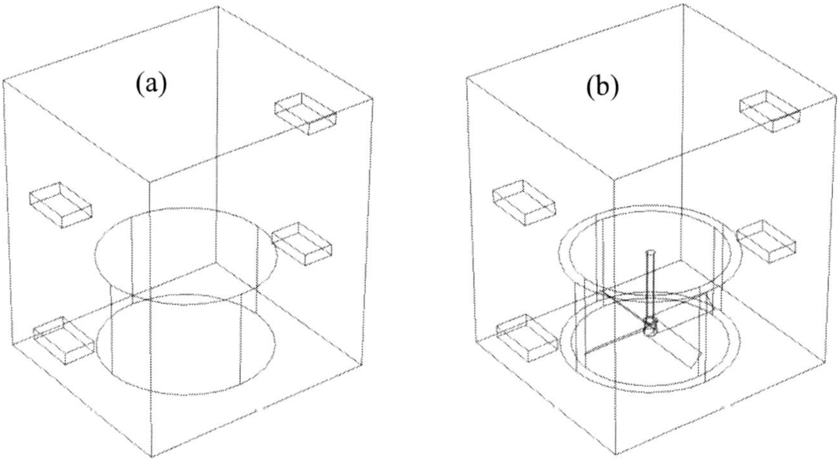

Figure 1. 3D geometries of microwave including the reactor (a) without and (b) with agitation.

$$\rho C_p \frac{\partial T}{\partial t} = \nabla(k\nabla T) - \rho C_p u \nabla T + Q_{mic} \tag{9}$$

where ρ is the density, C_p the specific heat capacity, T the temperature, t the time, k the thermal conductivity and Q_{mic} (W/m^3) represents the volumetric heat generation. The second term on the right is canceled when the system does not include agitation.

In these conditions, the microscopic energy balance is considered valid (Equation 9) with the following boundary conditions:

$$t = 0 \qquad T(x, y, z) = T_{ini} \tag{10}$$

$$t > 0 \qquad -k\,\nabla T = h\,(T - T_a) \tag{11}$$

The moisture loss during heating was not considered in the present model, this assumption is valid for short times of process.

Momentum Transfer Model

With the aim of modeling the distribution of the fluid flow within the tank, as a result of the agitation, the equations of continuity and the momentum transfer for laminar flow will be considered:

$$\rho \nabla u = 0 \tag{12}$$

$$\rho (u.\nabla) u = -\nabla . \left[-p + \mu \left(\nabla u + (\nabla u)^T \right) \right] + F \tag{13}$$

where u is the velocity, p the pressure, μ the dynamic viscosity and F the viscous dissipation term. In this case, the momentum transfer equation was solved in stationary state obtaining the velocities distribution in the entire volume.

To define the boundary conditions an external fixed domain and an internal rotating domain were established. The rotating domain is the internal subdomain and the rotating boundary condition was defined for the boundaries of blades and axis. The agitation velocity was set in 0.5 rps and the rotation direction was counter clockwise; and between the fixed and the rotating domains the boundaries must meet the continuity condition.

Numerical Solution

Tank without Agitation

The coupled energy and the electromagnetic field equations (Equations 1 and 9) with their initial and boundary conditions were numerically solved by the Finite Element Method (FEM), using the commercial software COMSOL Multiphysics.

COMSOL Multiphysics TM software uses a partition of the geometric model in simplest forms, for the application of the numeric method. The number of elements of the used mesh has influence on the model representation; in our work was solved with 51720 tetrahedral elements, using a differential meshing of the domain considering different subdomains such as the oven interior, the waveguide and the food (Figure 2). Inside the product a maximum elements size of 6 mm was incorporated, regarding the radiation wavelength within.

The analysis of the harmonic propagation was considered to solve the Equation (1), while the microscopic energy balance (Equation 10) was solved by the energy transfer module in transient state. The FGMRES solver was used to obtain the solution of the electric field in steady-state, considering a computational time of 73 s. To transfer energy in transient state the PARDISO solver was used, along with the BDF resolution package (multi-stage delayed derivation formula) which took 169 s to reach the solution for 100 s of processing.

COMSOL software estimates the electric field values in each position of the oven and inside the food; the microwave energy absorbed by the food and predicts temperature profiles inside the food as function of time.

Tank with Agitation

In this case, besides the equations of electric field and heat transfer, were also solved the equations of momentum transfer (Equations 1, 9, 12 and 13) with their initial and boundary conditions using COMSOL Multiphysics commercial software. Considering the complexity of the multiphysic system, equations to obtain both the distribution of the electric field inside the oven and product, as the distribution of fluid flow velocities in the tank, were solved in steady-state. The heat transfer equation was solved in transient state, so that temperature increase may be considered as a function of time.

To achieve a consistency in the meshing which considers the different physics, were taken into account both the element size that considers the wavelength of the microwave in the air and in the product (inside the tank), as well as the dynamic fluid effect produced by the agitation inside the tank (Figure 2). Regarding these aspects a mesh constituted by 265483 tetrahedral elements was reached.

In the case of the stirred tank, the stationary solution of the velocities field was incorporated, using the GMRES solver taking 474 s to reach the solution. To achieve the electric field solution took 296 s and to solve the heat transfer in transient state 634 s.

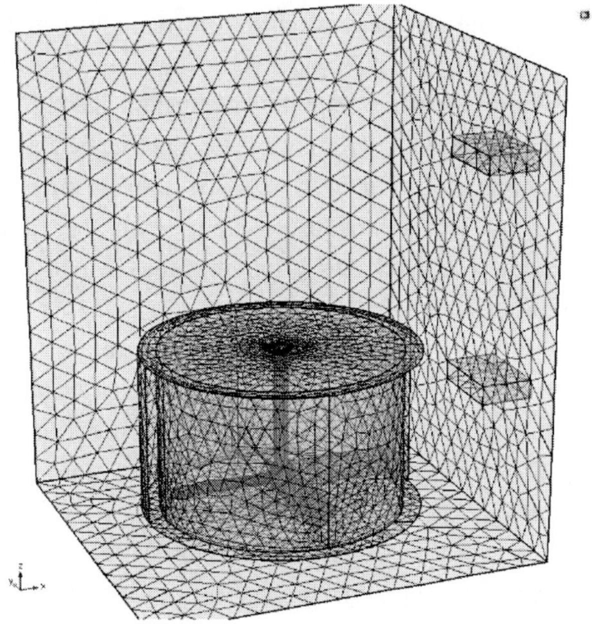

Figure 2. Domain meshing used for numerical simulations of the electromagnetic field, of the heat and dynamic fluid transfers using COMSOL Multiphysics software.

RESULTS

Table 1 shows the parameters and properties utilized by the mathematical model. They represent liquid products. The cavity dimensions (oven) are 38 x 35 x 43.6 cm.

Electric Field Distribution

Simulations of empty and loaded microwave oven were conducted. The numerical simulations allow discerning the displayed patterns when a load is placed in its interior. Figure (3) shows the distributions of the electric field intensity within the empty reactor. The software allowed observing the behavior of each component, in the x direction presents a nodal central plane with two planes with symmetric peaks and valleys. The component y is null in all analyzed planes, due to the waves negative interaction in this direction.

In component z, a resulting electric field distribution beneficial to the problem is presented, as shown in Figure (4a). It may be observed that the electric field intensity (norm) in the middle sector presents maximum values with a total of 12 peaks uniformly distributed. The distribution inside the oven is symmetric and these maxima diminish to the roof and to the floor. Figure (4b) shows the load effect on the electromagnetic profiles; the load distorts the maximums position, taking them to the top of the oven being asymmetric in the entire volume. Likewise, the maximum intensity of the electric field decreases at placing the load, mainly due to the dielectric capacity of the material, capable to dispel electromagnetic energy into thermal energy (Datta and Anantheswaran, 2005). It should be noted that temperature distributions within the load follow the electromagnetic field behavior (Campañone et al., 2014). Simulated results of temperature profiles for different arrangements are shown in the following analysis.

Table 21. Parameters and properties utilized in the mathematical model

Parameters	**Values**
Initial temperatures, °C	20
Initial velocity, m/s	0
Dielectric constant (ε’), dimensionless	9
Loss factor (ε”), dimensionless	3.2
Density, kg/m^3	972
Viscosity, Pa.s	1
Thermal conductivity, W/(m°C)	0.3
Heat capacity, J/(kg°C)	1800
Microwave frequency, GHz	2.45

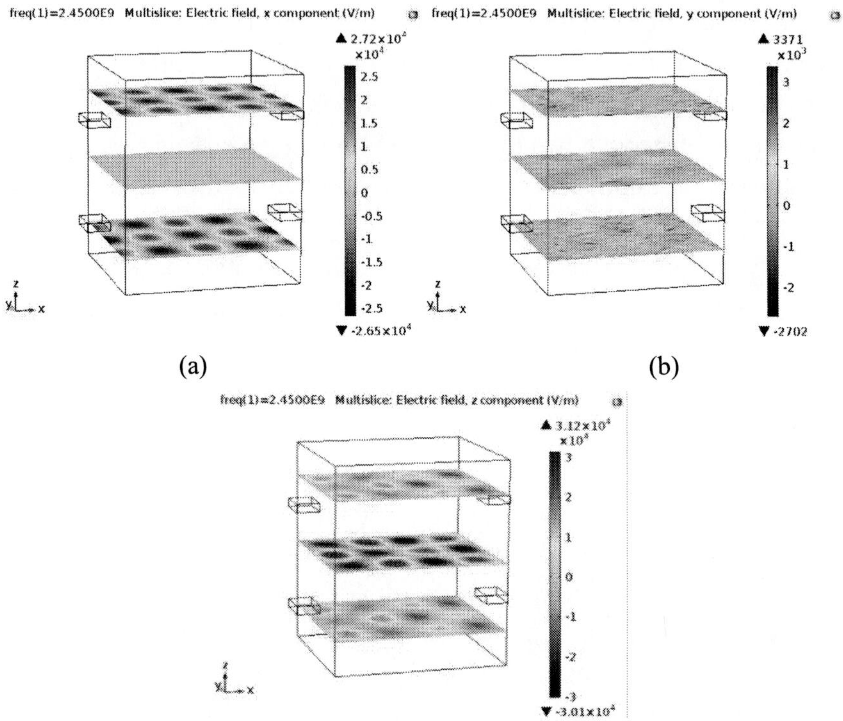

(c)

Figure 3. Electric field distributions for components in the reactor interior: (a) Ex, (b) Ey and (c) Ez.

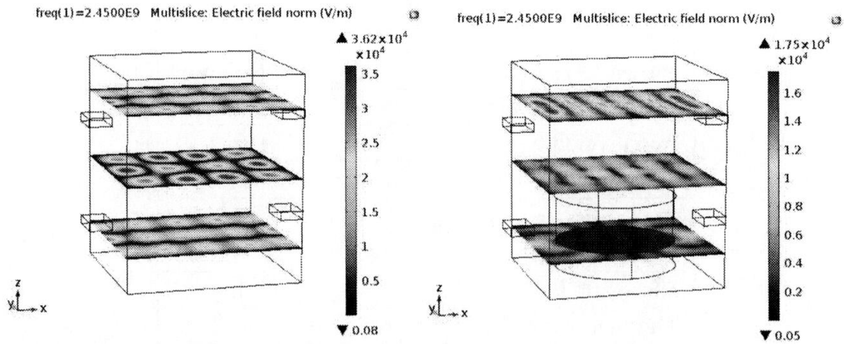

(a) (b)

Figure 4. Distributions of the electric field intensity in the reactor interior: (a) empty and (b) loaded.

Number and Position of Magnetrons

One of the objectives of the liquids thermal processing is to reach a uniform temperature throughout the product volume. For this purpose, different configurations with 1, 2 and 4 magnetrons of 400W power each were tested.

Figures (5-7) show the results obtained for 50 seconds of microwave processing. Figures show the temperatures distribution obtained at final time. Figure (5a) shows sample side heating when subjected to a single magnetron radiation; a progress in the upper side is achieved when adding another magnetron (Figure 5b).

To observe symmetry in heating the configuration was changed to two opposing magnetrons (Figure 6). This setting improved sample heating such as the final temperature value reached; however, the central area remains cold at the end of the process. A change in the magnetrons height (Figure 6b) did not show an improvement in the distribution nor at the final thermal level.

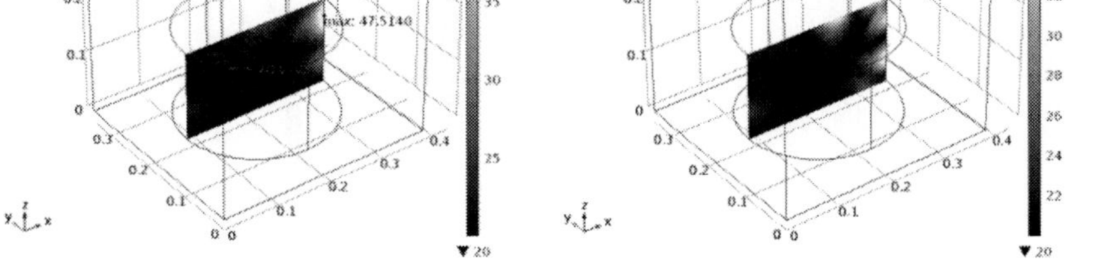

(a) (b)

Figure 5. Temperatures distribution, side heating (a-one and b-two magnetrons).

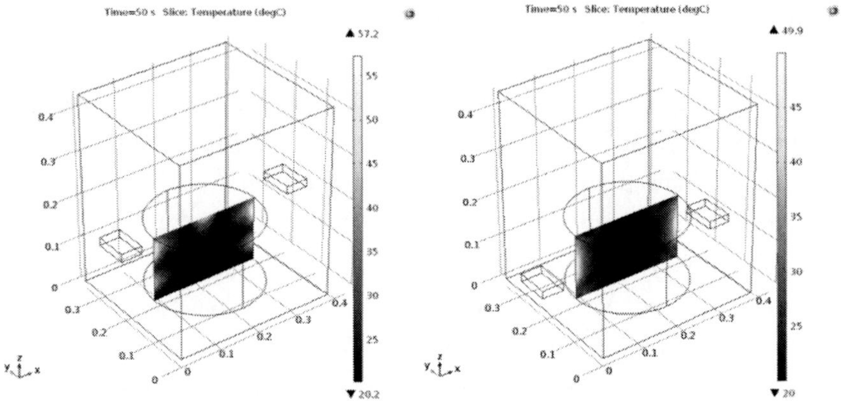

(a) (b)

Figure 6. Temperatures distribution, effect of magnetrons height (two opposing magnetrons).

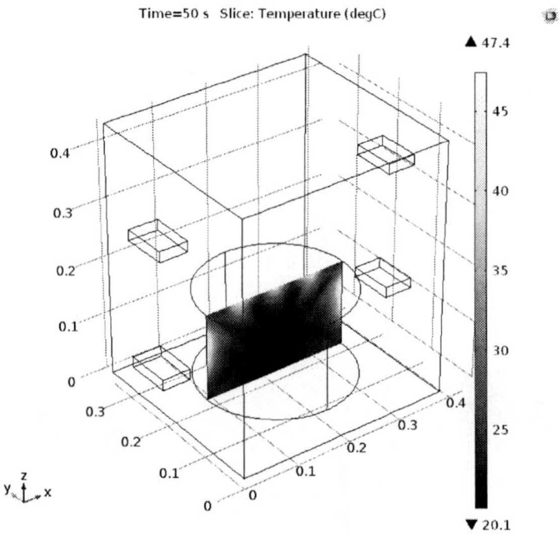

Figure 7. Temperatures distribution with four magnetrons.

Finally, the configuration was analyzed with 4 magnetrons (Figure 7). This configuration provided a more uniform volumetric heating. For economic and energy efficiency reasons is not suitable to continue adding magnetrons to improve heating, thus the incorporation of the motion of the control volume is used.

Designs Comparison

The reactor system without and with agitation was analyzed (Figures 1a and 1b). The microwave irradiation was conducted through 4 magnetrons of 1000W each, which are set as shown in Figure (7). Figure (8) shows the velocity profiles obtained through the resolution of the microscopic momentum balance along with the continuity equation. It may be observed that the blades motion causes a desired mixing effect; the velocity is maximal in the contact area with the blades and decreases in areas away from them.

The temperatures profiles in three positions inside the tank (lower, middle and upper area) are shown below and the results of the tank without and with agitation for 10 to 42s processing are compared (Figures 9, 10 and 11, respectively).

In the lower area (Figure 9) a better temperature distribution at 42 s heating in the stirred tank is obtained. In the case of the stagnant tank the pattern is preserved, and the cold areas remain both in position as in thermal level. The same occurs in the middle and upper area of the load. Due to the agitation, the temperatures increase in the upper area was greatly favored, where the stirred tank reaches higher temperatures and a better distribution than the tank without agitation.

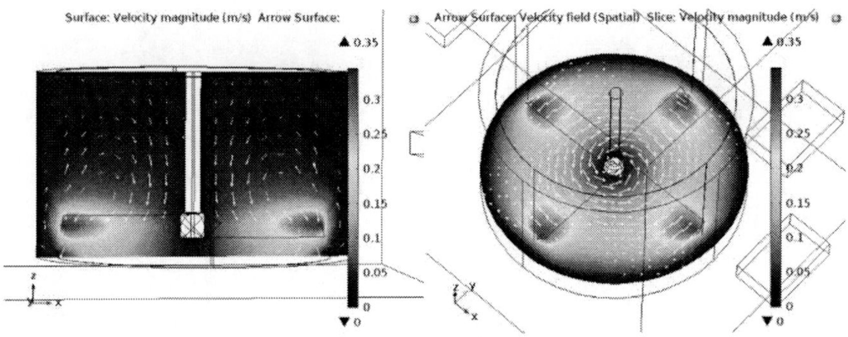

(a) (b)

Figure 8. Effect of the agitation on the velocities profiles: (a) front cut, (b) top view.

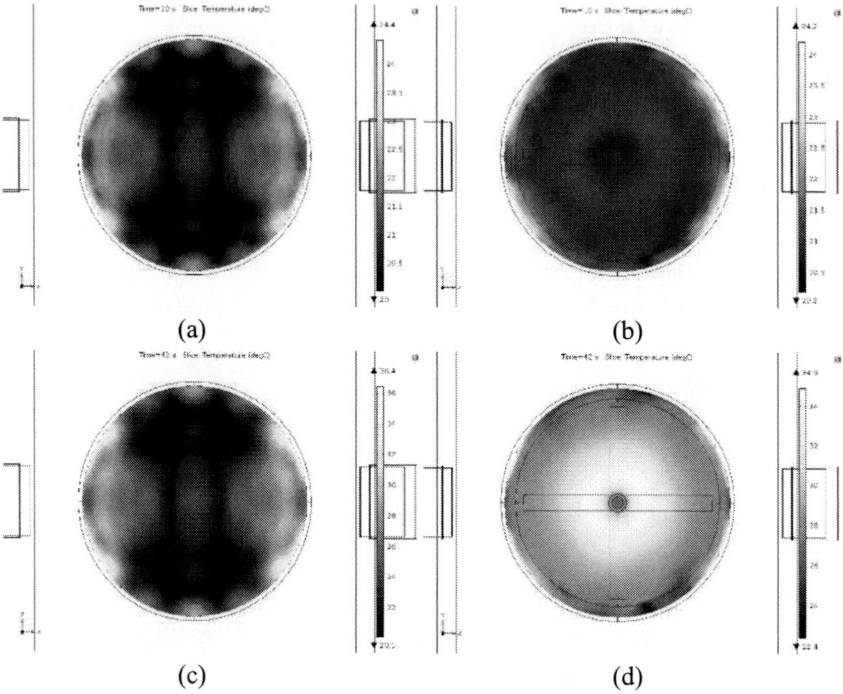

Figure 9. Temperatures distribution in tank without (a, c) and with agitation (b, d), lower area.

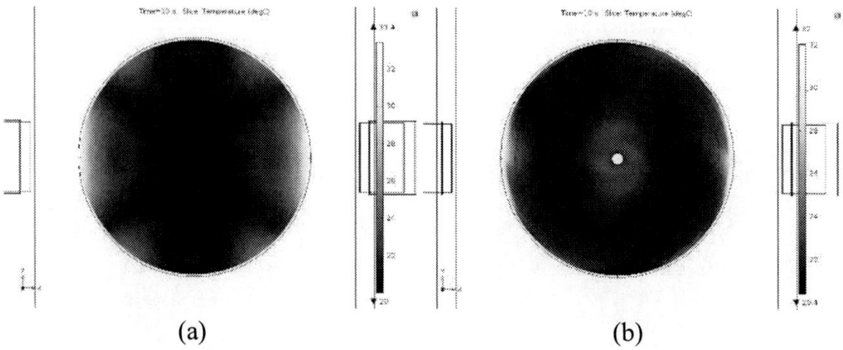

Figure 10. (Continued).

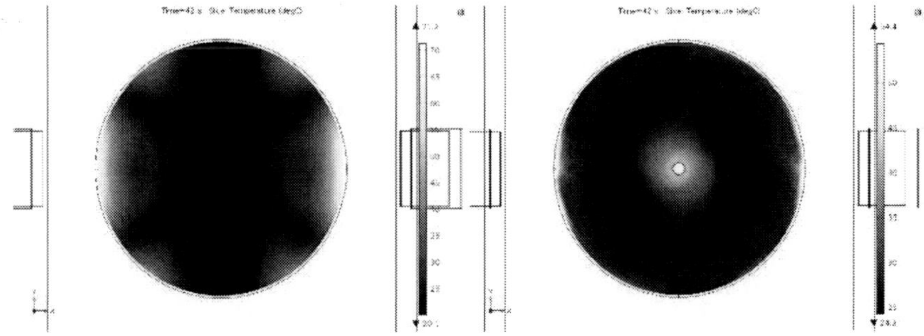

(c) (d)

Figure 10. Temperatures distribution in tank without (a, c) and with agitation (b, d), middle area.

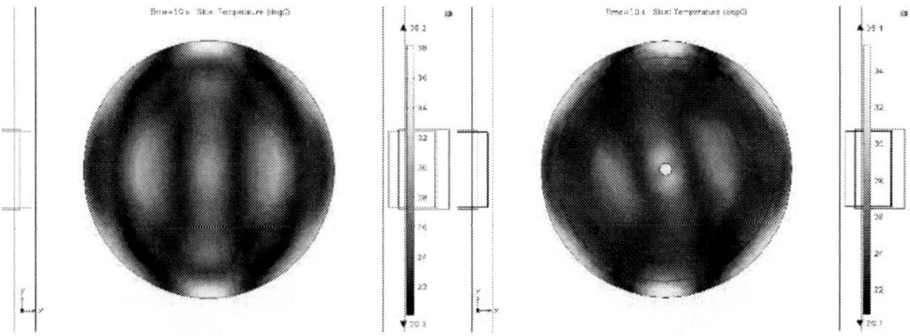

(a) (b)

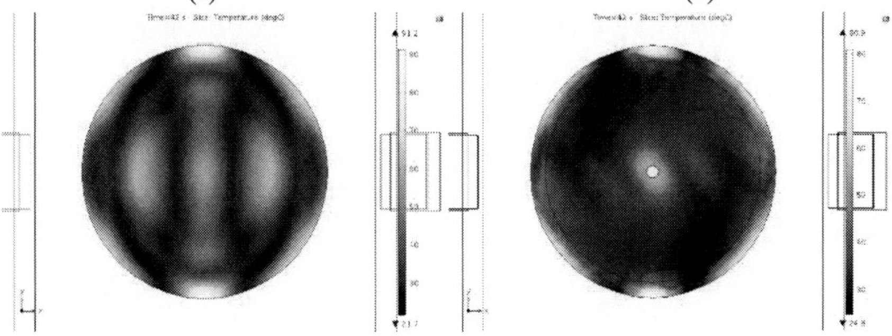

(c) (d)

Figure 11. Temperatures distribution in tank without (a, c) and with agitation (b, d), upper area.

One of the determining parameters for uniformity detection in profiles is the standard deviation of temperatures (Table 2). In the case of the stirred tank a 25% lower value than the stagnant tank was obtained, which would indicate that this technology is suitable to uniform heating using microwaves.

Table 2. Comparison of systems without or with agitation

System	Time(s)	Deviation
without agitation	42s	8.98
with agitation	42s	6.68

CONCLUSION

The present work addressed the resolution of coupled energy balance and momentum transfer considering valid Maxwell equations during microwave heating. Numerical simulations allowed to analyze different process scenarios such as the distribution and quantity of magnetrons as well as the effect of agitation on such systems. One of the conclusions of this work is the need for a great quantity of magnetrons to achieve heating in a stagnant system, which makes it inefficient. From this observation, and as a general conclusion of the conducted study, the use of agitation is vital to improve uniformity in temperature profiles. Finally, studies on this topic will intend to optimize the quantity, distribution and shape of the agitation systems in the levels and temperature distribution.

ACKNOWLEDGMENTS

The authors would like to thank the Faculty of Engineering, National University of La Plata (UNLP), and National Scientific and Technical Research Council (CONICET) of Argentina for the financial support.

References

Arballo, JR; Campañone, LA; Mascheroni, RH. Modeling of Microwave Drying of Fruits. *Drying Technology*, 2010, 28, 1178-1184.

Campañone, LA; Paola, CA; Mascheroni, RH. Modelling and simulation of heat and mass transfer during the heating of foods in microwave ovens under different heating schedules. *Food and Bioprocess Technology*, 2012, 5, 738-749.

Campañone, LA; Bava, JA; Mascheroni, RH. Modeling and process simulation of controlled microwave heating of foods by using of the resonance phenomenon. *Applied Thermal Engineering*, 2014, 73, 914-923.

Clemens, J; Saltiel, C. Numerical modeling of materials processing in microwave furnaces. *International Journal of Heat and Mass Transfer*, 1996, 39, 1665-1675.

Datta, AK; Anantheswaran, RC. *Handbook of microwave technology for food applications.* Marcel Dekker Inc. USA, 2005.

Dimian, AC; Bildea, CS; Kiss, AA. *Integrated Design and Simulation of Chemical Processes*, Elsevier, 2014, Vol. 35.

Dinčov, DD; Parrott, KA; Pericleous, KA. Heat and mass transfer in two-phase porous materials under intensive microwave heating. *Journal Food Engineering*, 2004, 65, 403-412.

Lin, YE; Anantheswaran, RC; Puri, VM. Finite element analysis of microwave heating of solid foods. *Journal of Food Engineering*, 1995, 25, 85-112.

Maloney, N; Harrison, M. Advanced Heating Technologies for Food Processing, Innovation and Future Trends in Food Manufacturing and Supply Chain Technologies. Chapter 8, Woodhead Publishing Series in *Food Science, Technology and Nutrition*, 2016.

Puri, VM; Anantheswaran, RC. Finite element method in food processing-a review. *Journal of Food Engineering*, 1993, 19, 247-274.

Verboven, P; Datta, AK; Anh, NT; Scheerlinck, N; Nicolai, BM. Computation of airflow effects on heat and mass transfer in a microwave oven. *Journal of Food Engineering*, 2003, 59, 181–190.

INDEX

D

E

F

G

H

I

K

L

M

N

O

P

Q

R

S